# Sustainable Waste Management and Energy Valorization - Challenges and Technological Innovations

*Edited by Hosam M. Saleh*
*and Amal I. Hassan*

Published in London, United Kingdom

Sustainable Waste Management and Energy Valorization - Challenges and Technological Innovations
http://dx.doi.org/10.5772/intechopen.1008018
Edited by Hosam M. Saleh and Amal I. Hassan

Contributors
Asma Nouira, Bibi Sadia, Diogo Canavarro, Faten Zubair Filimban, Gianni Innocenti, Gianpaolo Oteri, Huma Qureshi, Imen Amri, Imene Bekri-Abbes, Irene Mellone, Joel Sotamenou, Josephine Treacy, Mamoona Munir, Marco Frediani, Maria Ameen, Muhammad Faisal Amin, Muhammad Imran, Muhammad Iqbal, Muhammad Ravish Khan, Muhammad Saad Amin, Muhammad Zafar, Mushtaq Ahmad, Nabila Mahmood, Piero Frediani, Razia Tahseen, Roberto Simoni, Salman Majeed, Shabir Ahmad, Wajia Noor



First published in London, United Kingdom, 2026 by IntechOpen
IntechOpen is the global imprint of INTECHOPEN LIMITED, registered in England and Wales, registration number: 11086078, 167-169 Great Portland Street, London, W1W 5PF, United Kingdom

For EU product safety concerns: IN TECH d.o.o., Prolaz Marije Krucifikse Kozulić 3, 51000 Rijeka, Croatia, info@intechopen.com or visit our website at intechopen.com.

British Library Cataloguing-in-Publication Data
A catalogue record for this book is available from the British Library

Sustainable Waste Management and Energy Valorization - Challenges and Technological Innovations
Edited by Hosam M. Saleh and Amal I. Hassan
p. cm.
Print ISBN 978-0-85466-747-5
Online ISBN 978-0-85466-746-8
eBook (PDF) ISBN 978-0-85466-754-3

If disposing of this product, please recycle the paper responsibly.

# Meet the editors

Hosam Saleh is Professor of Radioactive Waste Management at the Radioisotope Department, Atomic Energy Authority, Egypt. He was awarded an MSc and a Ph.D. in Physical Chemistry from Cairo University. Saleh has more than 28 years of experience in hazardous waste management, with an emphasis on treatment and the development of new matrices for immobilizing these wastes. He is also interested in studying innovative, environmentally friendly techniques for managing hazardous and radioactive waste. He authored many peer-reviewed scientific papers and chapters and served as an editor of several books. He has been selected among the top 2% of scientists in the world, according to Stanford University reports published by Elsevier for 2019, 2020, 2021, 2022, and 2023.

Amal I. Hassan is Professor of Animal Physiology, Department of Radioisotopes, Nuclear Research Center, Atomic Energy Authority, Egypt. She has authored many peer-reviewed papers on chronic disease problems. She is a referee, reviewer, and editorial board member for several international scientific journals. She received a Certificate of Excellence in international scientific research arbitration from Publons. She was also selected for inclusion in Who's Who in 2014, 2015, and 2016.

# Contents

# Preface

Management of hazardous waste is one of the most critical and complex environmental issues of our age. It sits at the intersection of the development of industry, the health of the public, environmental stewardship, and technology. Whether in the cutting-edge laboratories of research institutions or in the highly regimented procedures of treatment, the effort to safely neutralize, destroy, or immobilize hazardous materials is a dynamic process. This book, *Sustainable Waste Management and Energy Valorization – Challenges and Technological Innovations*, is the result of a concerted effort to distill the active and adaptive landscape of hazardous waste treatment into a bold, forward-thinking document.

The vestiges of past industrial practice serve as a stark reminder of the damage that inaction can cause. At the same time, the unending development of waste streams, from e-waste to new chemical compounds, compels innovative and even more complex approaches. As the regulatory environment in many parts of the world continues to tighten, industries have moved not only to comply with regulations but also to adopt a philosophy of pollution prevention and a circular economy.

This book is intended to provide a thorough review of the state of the art and will consider the fundamentals of hazardous waste characterization, risk assessment, and the regulatory framework governing its management. The body of the book discusses a systematic inventory of treatment technologies. We take time to analyze familiar techniques such as solidification/stabilization, chemical treatment, and thermal destruction. However, we also allocate ample attention to biological techniques, newer physicochemical techniques, and emerging trends in bioremediation and phytoremediation. Throughout, the emphasis is not only on the technical workings of each technology but also on its use context, efficiency, constraints, and sustainability.

However, this is more than a technical manual. It refers to "current themes" by highlighting current problems, including the treatment of complex mixed wastes, the application of life-cycle assessment in technology selection, the potential for digitalization and automation in treatment facilities, and the economic drivers that both challenge and present opportunities for sustainable waste management. It also aims to look ahead to new or nascent technologies and research directions that are likely to change the field's relevance over the next few decades.

The chapters included in this work were written by an international group of experts, including academics, practitioners, and consultants, and we are proud to include persons from a multitude of backgrounds and experiences. By including a diverse group of collaborating authors, we have additional assurance of rigor and of grounding scientific concepts in a practical reality. We hope that this book will be a valuable resource for students starting a career in environmental management, for practitioners seeking to refresh their knowledge, and for decision-makers who will develop policies for the future of us all. Managing hazardous waste is a serious responsibility, a tribute to our commitment towards protecting human health and the life on the planet for the future of our children and grandchildren.

**Hosam Saleh and Amal Hassan**
Egyptian Atomic Energy Authority,
Cairo, Egypt

Section 1

# Systemic Challenges in Waste Management

Chapter 1

# Challenges of Domestic Hazard Wastes, with a Focus on Disposal, Treatment and Regulation

*Josephine Treacy*


## Abstract

Domestic hazard waste can be considered a hidden challenge to the environment. Domestic hazard waste can include home cleaning products, batteries, medicines, insecticides and herbicides, self-care products, and home maintenance products. Unfortunately, many of these products are unregulated in relation to domestic use. This chapter examines the uses of materials that can give rise to potential hazard wastes. The receptors after the use of domestic hazard waste and the effects the wastes can have on the environment and human health will also be discussed. The present disposal options and treatments will be outlined and gaps in knowledge surrounding regulation and treatment will be investigated. The outcome of this chapter will create awareness about domestic hazard waste and inform policy makers on the need for more regulation and management of domestic hazard wastes to enhance the attainment of the sustainable development goals.


**Keywords:** domestic, hazardous waste, receptors, treatment, regulation

## 1. Introduction

Waste management is a very important concern for all parts of society. Improper management of waste can affect environmental habitats and the overall quality of air water and soil ecosystems. The improper management can also have major impacts on human health and wellbeing. The focus of this chapter is to examine domestic hazard waste.

Unfortunately, in recent years, developed and advanced countries have experienced increases in living standards giving rise to ecological pressure on the environment [1]. As highlighted in the scientific literature, economic development and over consumerism [2] are associated with an increase in waste generation, which in turn contributes to climate change and the less availability of natural resources [2].

In relation to a European perspective, hazard waste is described as; *when it displays one or more of the hazardous properties listed in Annex III of the Commission Regulation No. 1357/20144* [3].

Hazard waste is characterized by the presence of a biological, chemical or physical property or a combination of these which are deemed hazardous. These characteristics can give rise to one or a combination of toxic, reactive, ignitable, corrosive, infectious and radioactive effects. Another important concept of hazard waste is synergy effects, which involves one component interacting with another with the introduction of a hazardous effect.

Unfortunately, household hazard waste is a global issue, the global trends of waste generation showed projections with an increase up to 2025 [4, 5].

Weak Waste Management can have major impacts on environmental habitats and human health [6]. The importance of understanding hazard waste and the challenges of disposal, treatment and regulation is the focus of this chapter. The production of waste is difficult to eliminate completely. Engagement with stakeholders and development of educational awareness and outreach approaches on the disposal, treatment and toxic effects on the biotic and abiotic components of ecosystem is important. This hopefully will be an mitigation measure, as society will be aware of the importance of more control on the use and treatment options of hazardous material.

Challenges related to the treatment, disposal and reuse utilizing modern technologies and encouraging innovations in reuses and treatments and disposal technologies are important to address. Comparisons of the quantities of hazard waste produced in different countries is difficult as globally countries classify hazard waste differently [7]. It is interesting to note, the presence of waste can create different levels of risk [8].

$$\text{Risk} = \text{Hazard X Exposure} \qquad (1)$$

When discussing waste, waste treatment and management, it is all about controlling the risk. The risks can involve, for example, pathological or infectious and harm such as a chemical burn. The risks can be immediate as well as effects from prolonged exposure.

Technology for treatment can include chemical, thermal, biological and physical processes or a combination of these approaches. These approaches can include recycling, composting, waste incineration, mechanical-biological treatment and landfilling. More advanced treatments such as supercritical water oxidation, microbial bioremediation and electrochemical processes can also be utilized. Policy surrounding domestic waste and domestic hazard waste including transboundary effects in a European context is important to study as it can guide treatment and management strategies for hazardous waste [9].

Domestic hazard waste can be a hidden challenge as it is hard to regulate, especially in relation to private dwellings. The public might not be aware of the hazardous nature of the waste and what treatments are needed. This chapter hopes to address some of these issues.

## 2. What is hazard waste?

Hazard waste can be found within the realm of waste. Hazardous waste poses a risk to human health and the environment. Hazardous wastes should not be handled and disposed of through the non-hazardous waste streams process. An interesting survey from 2022 shows the different categories of waste (**Figure 1**).

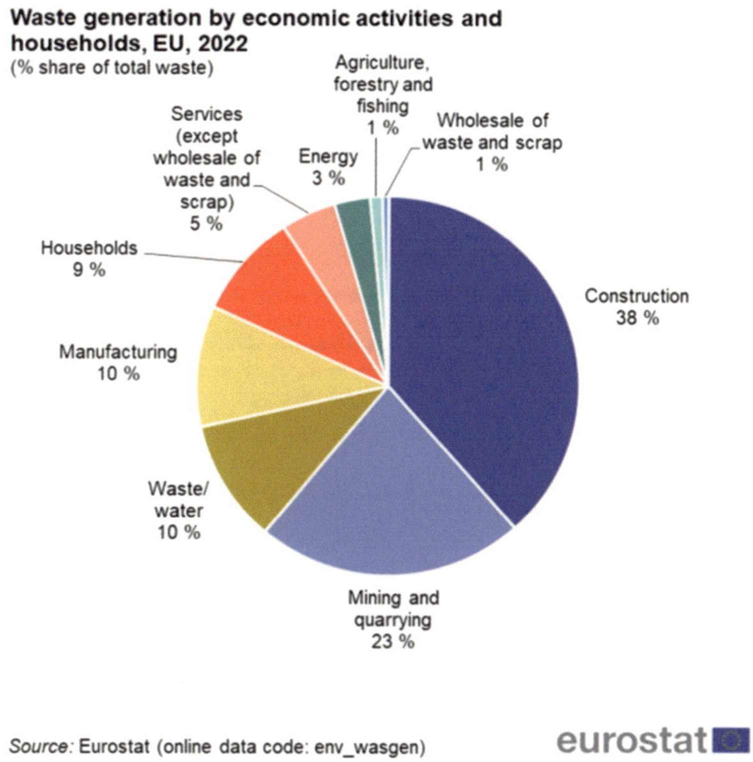


**Figure 1.**
*Waste generation types from an EU study in 2022 [10].*

The classification of hazardous waste can be based on the following explosive, oxidizing agent, flammable waste, skin irritation and eye damage, specific target organ toxicity (STOT), aspiration toxicity, acute toxicity, carcinogenic waste and infectious waste. Others include mutagenic waste that may cause a mutation of the genetic material in a cell, waste that releases acute toxic gases in contact with water or an acid, eco-toxic waste that may cause immediate or delayed risks for one or more sectors of the environment and waste capable of exhibiting a hazardous property not directly displayed by the original waste (**Figure 2**).

## 2.1 Seasonal trends and variations per country

Seasonal trends concerning the type and the volume of waste in use should be taken into consideration when assessing the environmental impacts of hazardous waste at different times of the year and the level of management needed. A study in a Copenhagen suburb examined seasonal waste trends within Autumn, Winter and Summer in the year 2011–2012. The findings found the highest volumes of waste were generated in the summer period [11].

A world bank report shows the variation of waste produced in different parts of the world which is graphically presented in **Figure 3.**

A study from Canada found 39% of households had expired medicines to dispose of in their homes, 58% of households had unwanted batteries of which 42% were placed in general waste [12]. Another study evaluated the generation and types of

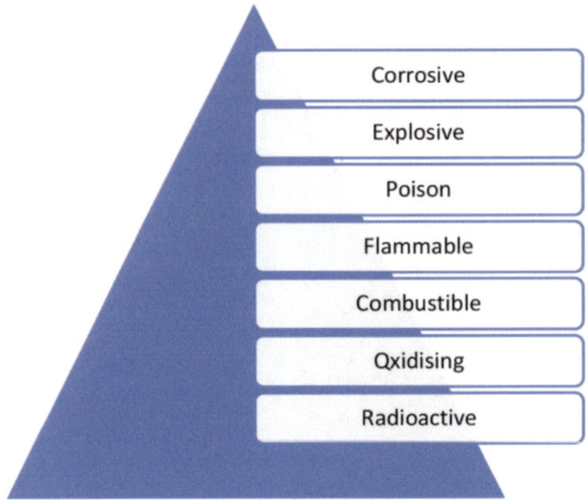


**Figure 2.**
*Saliant words describing forms of hazardous waste.*

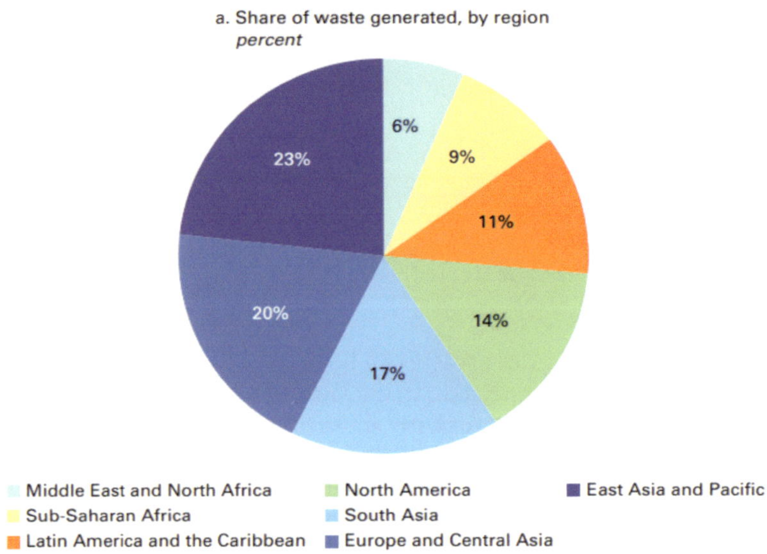


**Figure 3.**
*Percentage of waste generated in different parts of the world [5].*

household hazard waste in Nakhon Si Thailand by questionnaire design and distribution. The questionnaire results revealed the amount of hazard waste per household was 11.95 kg per year of which 17.4% was electronic waste, 16.98% was fluorescent lamps and aerosol spray related to 12.38% and cleaning products made up 11.3% while engine and lubricant oil were found to be 9.87% [13].

## 2.2 Sources and receptors of hazard waste

Root cause and reduction at source is always recommended. An interesting study shows the importance of source reduction due to the costs involved in managing hazard wastes from a global viewpoint [14]. Source, types and receptors of a selection of waste streams can be seen in **Figure 4**. When assessing pollution risk, it is important

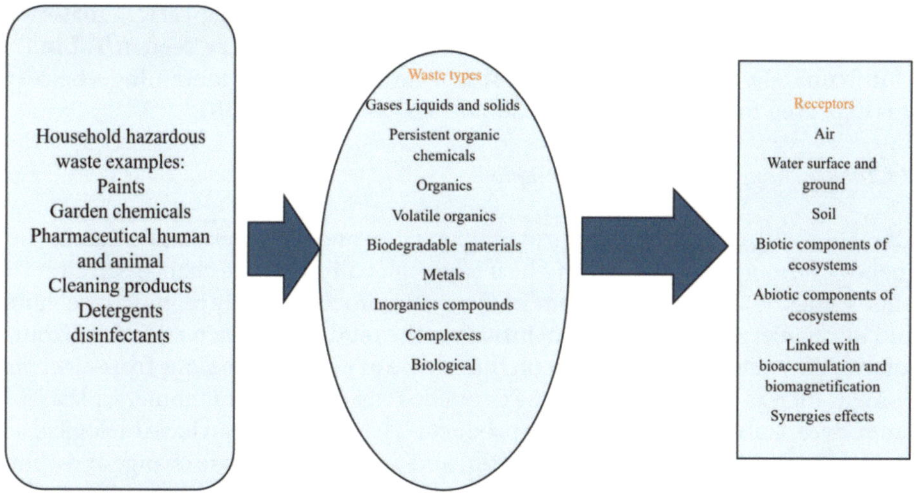


**Figure 4.**
*Source and receptors of hazard waste.*

to examine the sources, pathways and receptors within a model approach [15]. There can be multiple pathways with various receptors. The United Nations environmental management group have discussed multiple pathways and multiple receptors in line with understanding the exposure and risks related to pollutants [16].

Household hazard waste can consist of toxins such as organics, heavy metals and nutrients. The public, when handling hazard waste materials due to their lack

| Product | Hazard |
|---|---|
| Pet flea powders | Chlordane |
| Paint strippers Paints, adhesives, solvents | Methylene Chloride |
| Shoe polish | Nitrobenzene |
| Rodent control | Warfarin |
| Insect repellents example aerosols | Hexachloroethane |
| Waste batteries and household batteries | Heavy metals including lead |
| Waste electrical and electronic equipment (WEEE) Electrical and electronic equipment includes household appliances, telecommunication devices, computers, office equipment, telephones, cameras, radios, lighting equipment, electrical tools, toys with electrical or electronic components, and other automatic devices. | metals |
| WEE Plastics<br>Pharmaceutical wastes [17]<br>Waste waters: | Polystyrene (42%), acrylonitrile-butadiene-styrene copolymer (38%) and polypropylene (10%) were found in WEEE plastics<br>Example: Cholesterol drugs such as atorvastatin<br>Gray water and black water and sewage: Metals, surfactants bacteria residual pharmaceuticals. |

**Table 1.**
*Interesting domestic hazards.*

of knowledge, might not understand the hazard characteristics of certain materials. Common hazard waste that the public regularly encounter can be seen in **Table 1**.

Unfortunately, the global e-waste from discarded products containing a battery or plug is expected to reach 74 Mt. by 2030 and 120 Mt. by 2050 [18].

*2.2.1 Climate change and pollution receptors*

Climate change can influence mean temperatures and sea-levels, and enhance the frequency of droughts, flooding, and heat waves globally. Climate change can alter the baseline values of abiotic components of the environment, namely temperature, humidity and ultraviolet radiation. This can influence the release of certain emissions from pollutants. An example is the impact on the release of certain emissions from electrical equipment such as WEE plastics. These emissions can consist of monomers, plasticizers, solvents, dyes, stabilizers, degradation products. The emissions can have biological activity and can contain heavy metals such as tin, and antimony. Climate change and climatic events such as flash flooding can increase the risk of more leakage of hazard waste from management sites [19]. The pathways could be soil and then water and depending on the volatility nature of the pathways could be soil and then water and depending on the volatility nature of the pollutant, the air can be the receptor, which can give rise to greenhouse gases and thus impact climate change [20]. Heat waves increase odour, dust and particulate matter from treatment sites. Particulate matter can deposit on soils and then enter rivers and lakes aided by the increase in precipitation due to climate change [21].

## 3. Disposal and treatment of waste

Production of domestic hazard waste streams can give rise to disposal and treatment needs. Hazard waste if not managed correctly can give rise to pollution concerns.

### 3.1 Treatment and challenges of hazardous waste

When reviewing management of waste options, the following variables are important to understand waste identification, minimization strategies, storage and handling, as well as treatment and disposal. Regarding hazard waste management, integrated waste management should be the forefront. Integrated waste management involves engaging in social, economic and environmental aspects. The preference of treatment should follow the approach of **Figure 5**.

The onus on society, stakeholders and innovators is to design products and packaging with the intention of reducing the quantity of use and the toxicity of raw materials [22]. A recent European regulation promotes ecodesign in manufacturing aims to reduce the source of hazard wastes [23]. This regulation is in-line with the United Nations, sustainable development goals especially with goal number 12 which relates to responsible consumption and production. The aim of efficient treatment for energy and resource recovery include: high-product quality raw materials and energy release potential after treatment [24].

### 3.2 Treatment approaches

As an initial protocol; segregation and control of volumes of hazardous waste will aid the treatment process. Examples of sustainable practices related to waste include

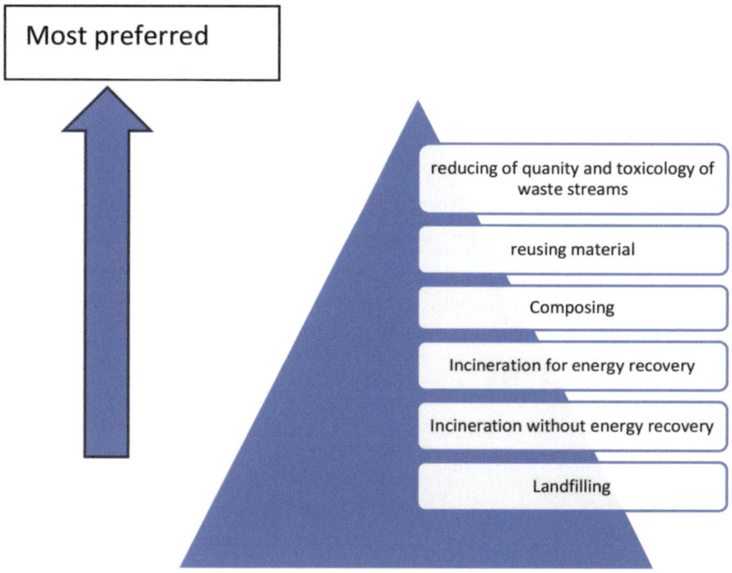


**Figure 5.**
*Order of preference management of waste.*

recycling, composting (difficult to achieve with hazard waste), incineration, pyrolysis and gasification. Treatment with the result of a new product or resource and energy recovery from the waste the most sustainable practice. Energy recovery can benefit sustainable development goal 7 which relates to affordable and clean energy and sustainable and development goal 13 which engages with climate action [25]. Energy recovery can also aid smart consumerism by moving away from the use of fossil fuels. Energy from waste can enable energy security of a country [26]. European countries engaging with energy from waste can be seen in (**Figure 6**) [27].

### *3.2.1 Recycling*

Plastic and electric waste and glass can be recycled if managed correctly [28]. Recycling is a clean approach to managing hazard waste. An application of glass recycling in concrete production and as a soil enhancer demonstrates innovative use of glass [29].

### *3.2.2 Composting*

Composting can involve an aerated static pile which forces ambient air through the compost pile. This method is commonly used for municipal solid waste at an industrial scale and generally requires less land area and relatively less time for the attainment mature composting. In relation to in-vessel composting approach, the composting is within a container or vessels. The loading capacity and performance of the composting process is limited by microbiological components. Aerobics, anaerobic and vermi processes can be utilised. Composting is not suitable for hazard waste, when composting waste such as food stakeholders need to be careful there is no contamination from hazardous materials such as herbicides in the compost mix [30]. The European compost network have highlighted the challenges of contamination of home compost with plastic residuals [31].

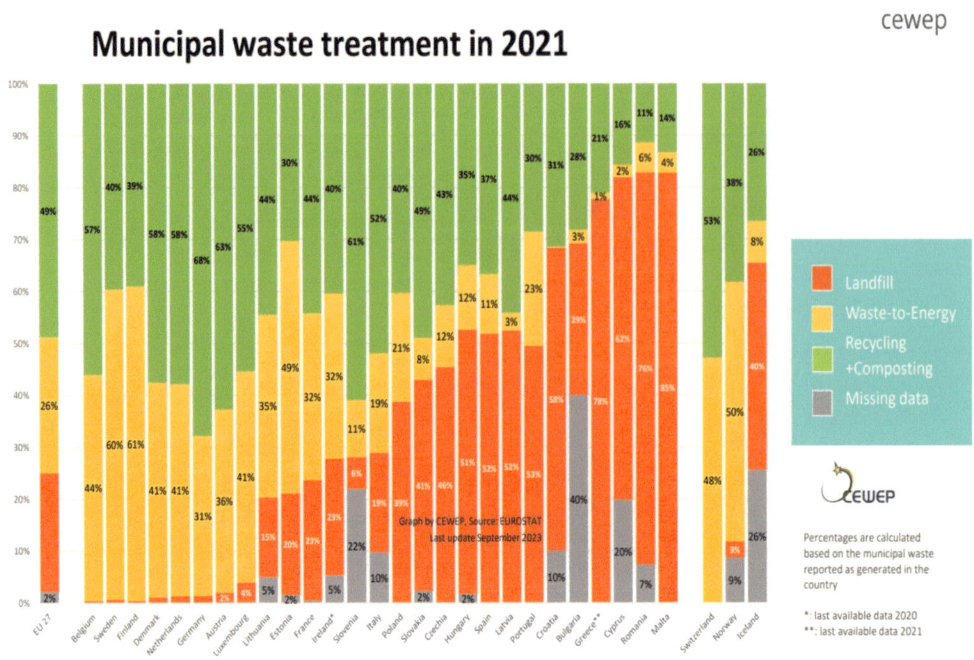


**Figure 6.**
*Types of treatment in different countries in Europe in 2022 [27].*

### *3.2.3 Landfill*

Landfill technology is not very desirable for hazardous waste due to environmental risks related to leakage from the landfill and emission issues. The hazard waste would need to be very stable or treated for stability before even considering a disposal option of landfill.

The Landfill Directive requires very strict regulation on the design of the landfill parameters such as the geological barrier. Site selection and understanding the risks related to ground water, air and soil quality must be fully understood [32]. Essential monitoring and maintenance of soil cover thickness, leachate collection and landfill gas recovery are important. The weather conditions, temperature and precipitation, pH of leakage and biodegradable matter loading are all important parameters affecting the emission of landfill gases and the toxicology of leachate [32].

The aim of the European landfill directive is to ensure progressive reduction in the landfilling of waste. Landfills must not accept flammable, explosive or corrosive wastes or wastes from hospitals and medical products by 2030 according to the European union directive. The EU directive also states that the share of municipal waste being landfilled be less than 10% by 2023 [32].

### *3.2.4 Waste water treatment plants and septic tanks*

Domestic liquid waste consists of water 99.9% and small amounts of solid inorganic and organic components. Key components consist of food waste, and human waste including nutrients, nitrogen, phosphorous and microorganisms including bacteria and viruses. The overall workings of a waste water treatment system is described in the literature [33]. Centralized wastewater treatment plants and septic tanks are important treatments for human discharge waste which is known as black water. Gray water, which relates to washing clothes and for example bathing in most cases passes through the above-mentioned treatment approaches. Other treatment systems consists of septic tanks, aerobic treatment units and constructed wetlands.

The onus for society is to be aware of what should not go down the kitchen sink or toilet [34]. Disposal of hazard material down a sewer can cause, blockage to pipes and upset the overall performance of the treatment process.

The sludge formed during treatment of water is techniques such as dewatering and digestion aerobic and anaerobic techniques. The methane released can be used as an energy source [33]. Liquid hazard waste other than sewage should be treated with care and should not be poured down drains. In European countries, advise is available to society on how to dispose, of this domestic liquid waste such as in public amenity centre [35]. Underdeveloped countries presently do not have the infrastructure to guide society on the proper treatment, more financial supports and coordinated of waste management practices are needed to meet present and future demands [36].

#### *3.2.5 Thermal treatments*

Thermal treatments have the advantage of destruction of the waste and having the option of energy recovery from the waste [37]. The alarming situation of using fossil fuels with increased emission of greenhouse gases, climate change, polluted environment, and increased health risks are a hindrance globally. Clean and renewable energy sources are in desperate need to solve the current issue and enhance sustainable living. Pyrolysis, gasification and incineration are the commonly used methods for waste to energy conversion.

##### *3.2.5.1 Pyrolysis*

Pyrolysis is a thermal degradation process of organic and hazard waste materials without the use of oxygen or air. The waste most be suitable for pyrolysis include, plastics, biomass, tires, medical waste and household waste [38]. Pyrolysis and co-pyrolysis have less emission of air pollutants and act as alternative options to incineration for hazardous materials [39].

An interesting pyrolysis plant is located in Germany company name Arcus, which processes 4000 tonnes of plastic waste per year [39]. The process occurs at high temperatures. The products created after pyrolysis include char, condensable liquids, light and heavy oils and tar. Techniques can be catalytic fast, intermediate, slow or vacuum processes. Commonly Pyrolysis is undertaken under atmospheric pressure. Vacuum pyrolysis can be carried out under low pressure. The vacuum technique has a lower residence time of organic vapor in the reactor and has a lower decomposition temperature [40].

##### *3.2.5.2 Gasification*

Pyrolysis can go through an additional process to produce a gasification product after the addition of water. Gasification is a thermochemical degradation process that converts carbonaceous components into gaseous form, predominantly $CO_2$ and $H_2$ gases [41]. Gasification is one of the most promising techniques to valorise waste material [42]. The process is carried out within a temperature range of 800–1200°C [42].

To date pyrolysis and gasification are not readily used in comparison to incineration. In relation to urban hazard wastes Japan have developed plants using pyrolysis. In the European union there is a large industrial facility in Burgau in Germany.

*3.2.5.2.1 Incineration*

Incineration has the ability to reduce the weight by about 70% and the volume by 90% [43]. The fly ash produced during incineration can be a pollutant if not managed correctly. The hazard nature of the fly ash is due to the presence of metals and organics. It can also be known as particulate matter PMx (x refers to the diameter of the particle). Separate treatment can be carried out on the fly ash such as stabilizing at temperature of 800°C which prevents leakage of metals. On a positive note, the fly ash has many resources which can be recovered such as metals.

The different types of incineration designs are rotary kiln, fluidized bed, moving grate, liquid injection and multiple hearth. Energy recovery makes the process sustainable [44]. Indaver a leader in waste management utilize incineration in the following manner,; *'Non-recyclable waste is converted into energy in grate incinerators and fluidised bed incinerators. Their rotary kilns are reserved for the safe thermal treatment of hazardous waste'* [24].

*3.2.6 Supercritical fluid water*

A supercritical fluid has a diffusability characteristic of a gas and the solubility characteristic of a liquid. Supercritical water is water that is above its critical temperature and pressure within its phase diagram. The critical temperature and pressure of water is 374.3 °C and 22.12 MPa. The supercritical phase can be spiked with oxygen and hydrogen enabling a supercritical water oxidation to occur. The breakdown of waste produces water and carbon dioxide [45]. A study of using supercritical fluid water oxidation for the treatment of pharmaceutical waste gave rise to a total organic carbon removal by 99.9% [46]. An interesting study of sludge from wastewater treatment, further treated using supercritical fluid water oxidation enabled the overall treatment process more efficient [47, 48].

Pilot research has shown the use of supercritical water oxidation, for the breakdown of organics such as polychlorinated biphenyls, phenol and landfill leachate. Challenges on the use of supercritical water oxidation technology include cost of operation and safety factors while operating at high temperatures and pressures including corrosive aspects. Research in making the technology more efficiency by improving the mentioned challenges is ongoing [49].

*3.2.6.1 Electrochemical*

Electrochemical techniques are not commonly used but they have a large potential role in the future. Techniques such as electrodialysis, electrochemical oxidation and electrochemical reduction can be used for mixed and hazard waste [50]. Electrochemical oxidation for example changes metals to a reactive valance state which move towards the pollutants speeding up the degradation [51].

Their deployments for wastewater management, depend on the volume of water, as well as the contaminant. The methods that can quickly adapted to Electrochemical processes have the advantages of being robust, having simple operational management, and being able to quickly adapt to wastewater changes by simply turning on and off the power and/or altering the current density [51].

As an overview, minimization and recycling are always the best option. In relation to thermal treatment, incineration is still the most proper choice where feasible. Biological treatment using microorganism to degrade organics in wastewater streams has a large role to play in domestic liquid waste. The importance of engaging with amenity sites for proper disposal of hazard solids and liquids is important to encourage within society. Amenities site approaches should be encouraged in all parts of the world. Risk factors for all management options are important to measure as even controlled disposal and engaging with treatment can also give rise to pollution concerns during the process; some examples can be seen below **Table 2**.

Innovations in technologies such as carbon capture and storage, CCS can make technologies more sustainable [52].

The world bank **Figure 7** have highlighted the percentage of different types of technologies used throughout the world.

The World Bank Highlights in **Figure 8** landfilling and thermal treatment of waste are the most common methods in high income countries. Several middle-income countries have poorly operated landfills. The report also states lower income countries dispose of waste in open dumps. When deciding on a waste treatment, emphasis on the contribution to the circular economy should be considered [53]. This is very difficult in certain countries of the world where services and financial support are not available.

| Technology | Environmental concerns if not optimized and managed efficiency |
|---|---|
| Landfill | Odors, green house gases, water leachate; metals, organics and pathogens. |
| Incineration | Atmospheric pollutants, fly ash, odor from example $SO_2$ NOx, $CO_2$ |
| composting | Leakage, green house gases |

**Table 2.**
*Examples of emissions from treatment technologies.*

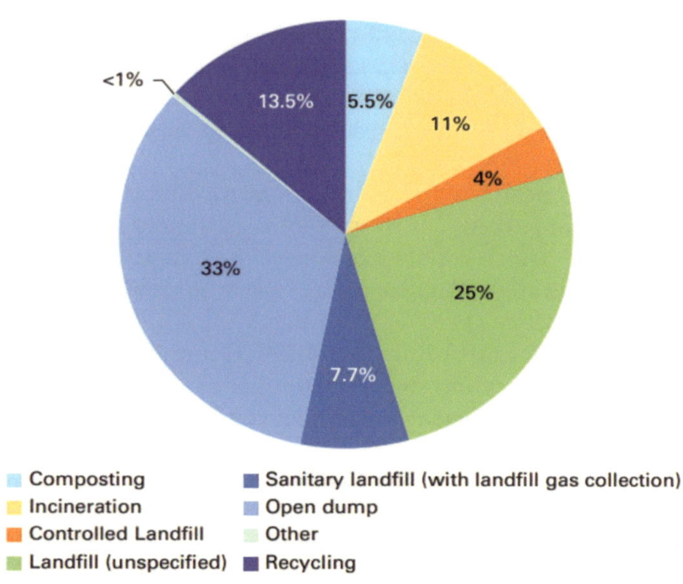


**Figure 7.**
*% of different technology of waste management used globally (Kaza et al. [5]).*

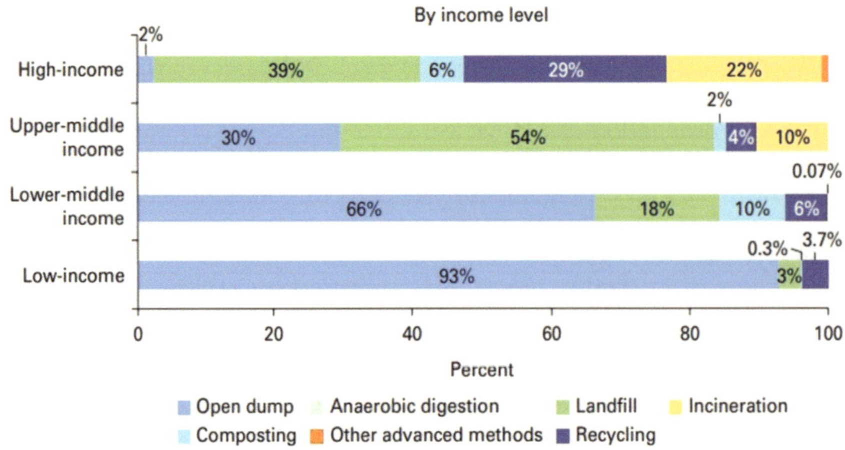


**Figure 8.**
*Variation in types of treatment in relation to income [5].*

## 4. Policy and regulation

Policy and regulation in relation to waste aims to promote prevention and efficient management. Other levels of policy and regulation aims to engage with preparation for reuse and recycling. Hazardous waste's policy and regulation, are difficult to audit fully within a domestic setting [54]. As domestic households have a certain level of self-control over their waste streams. Knowledge and education for stakeholders is so important where hazard waste is concerned [55–57].

Important legislation includes, the European waste framework directive [3], Directive 2008/98/EC the landfill directive [19]. The packaging and packaging waste PPWD and the single use plastic directive (SUP), which was first laid out in the Packaging and Packaging Waste Directive 94/62/EC (PPWD) and now the Packaging and Packaging Waste Regulation 2025/40 (PPWR) [58].

*The Regulations formally set out the following waste hierarchy which shall apply as a priority order in waste prevention and management, legislation and policy: (a) prevention; (b) preparation for re-use; (c) recycling; (d) other recovery (including energy recovery); and (e) disposal.*

In relation to the European Union Waste framework directive, a brief from the European Environmental bureau, (EEB) highlights an important change in relation to household hazard waste, which will improve recycling potential. From January 2025 it is obligatory to separate and collect hazardous waste as stated in article 20 as part of the waste framework directive (WFD) [59, 60].

Another important policy is the European REACH initiative, which relates to registration, evaluation, authorization and restriction of chemicals. This policy advises manufacturers in the use of less hazardous chemicals within their manufacturing process [61].

The Extended Polluter Pays responsibility initiative is hard to implement for domestic settings. A set of new guidelines has created more confidence approaches to solving problems due to the addition of 18 Producer Responsibilities [62].

Another important initiative which applied to household waste, is known as the Basel convention which controls the movement of hazard wastes between countries and addresses transboundary effects [9, 63, 64]. Since the 1980s the cost of toxic waste disposal is very expensive. The cheapest option for many high-income countries is to export to underdeveloped countries for treatment, such as African and Eastern European countries. As a consequence, waste can be improperly managed, in response to this, the Basel convention protects developing countries by regulating the trade of hazard waste [9].

A global trend assessment, the Global Monitoring (GMP) plan, aims to analysis and compare the values of persistent organic pollutants (POP) within global and regional areas. It helps to understand the trends in different parts of the world and aids understanding persistent organic pollutants transboundary effects. The Stockholm convention initiated the Global monitoring programme [5].

The circular economy and the concept of zero waste is important for the management of waste. The European Union (EU) follows the 10 R principles which relate to redesigning technology, repair, refrain, repurpose, reduce, remanufacture, re-mine, recover, recycle, and reuse. However, most theory surrounding the circular economy are based on the 3 Rs "reduce, reuse, recycle" [57, 65–67].

Other European legislation include the WEEE Regulation-Waste electrical and electronic equipment S.I. No. 340 of 2005 these regulations include WEEE return scheme. Another important regulation is the, European communities (waste oils) regulations S.I. No. 399 of 1992.

Another policy and regulations such as the: *Waste management (hazard waste) regulation S.I. No 163 of 1998 S.I. No. 73 of 2000; Waste Management (Movement of Hazardous Waste) regulation S.I. No. 147 of 1998; Waste Management (shipments of Waste)* Regulations 2007 S.I. No. 419 of 2007, demonstrates the microsphere and microsphere involvement in policy development and implementation.

The Hazard Waste regulation has prioritized the additional requirements for labelling, record keeping, monitoring and control obligations for the "cradle to the grave for the hazard waste. The directive also bans the mixing of hazardous waste with other categories of hazardous waste and with non-hazardous waste [68].

Performance and data surrounding policy is important to gather. A world report states waste management data is important to critically evaluate, for the understanding of policy and planning needs into the future [5].

Most countries globally view waste management as a local responsibility. In addition to this, globally 70% of countries have established policy and regulation to guide management.

## 5. Conclusions

This chapter outlines challenges in terms of treatment, and regulation of hazard waste. The chapters finding highlight, Hazard waste is challenging to manage, especially in relation to domestic waste due to the fact that domestic waste comes from private citizens. Prevention is always the best approach. Education and awareness on the use and disposal of hazard waste is important to encourage [69–72]. How household waste is handled, and disposal of though civic amenities, should be part of any educational programme [69]. Recycling waste is the best option where viable [59, 73, 74].

The risks of contamination when recycling must be accessed when deciding on the technology being utilized.

All treatments can give rise to pollution emissions, so it is important to control emissions when treating hazard waste [75].

The most popular treatment is incineration. Other innovative treatments namely pyrolysis, gasification, supercritical fluid oxidation and electrochemical have a place in the future.

Treatment that can produce a fuel as an alternative to fossil fuels is already going to benefit society and of course the sustainable development goals [76]. As stated in chapter energy security will always be a driver to hazard waste treatment.

Where managing product design, it is all about reducing the toxicology of the product where possible and to be innovative in the produce's manufacturing process to engage with green manufacturing [77]. The encouragement of the use of green labels on products would benefit the minimization of the use of hazard materials [57, 78]. The green label was discussed under policy initiatives. The environment needs to be protected as well as human health into the future. Continuous emphasis on education and innovations in technologies for treating hazard waste is important to nurture. The concept of zero waste is always important or moving in the direction towards zero waste aided by policy and regulation. Local waste stewardship initiatives should also be encouraged to minimize the challenges of managing hazard wastes [66].

Global harmonization of technologies is important to envisage, in order to ensure low income countries and developing countries have equal opportunities to utilize the best technologies to treat their wastes.

Ethics within waste management, the sustainable development goals, communities and citizen science involvement in planning, should always be the salient considerations when designing policy and managing waste into the future.

## Conflict of interest

No conflicts.

## Author details

Josephine Treacy
Technological University of the Shannon Midlands Midwest, Moylish Campus, Limerick, Ireland

*Address all correspondence to: josephine.treacy@tus.ie

## References

[1] He H, Reynolds CJ, Zhou Z, Wang Y, Boland J. Changes of waste generation in Australia: Insights from structural decomposition analysis. Waste Management. 2019;**83**:142-150. DOI: 10.1016/j.wasman.2018.11.004, ISSN 0956-053X

[2] Minelgaitė A, Liobikienė G. Waste problem in European Union and its influence on waste management behaviours. Science of the Total Environment. 2019;**667**(2019):86-93. DOI: 10.1016/j.scitotenv.2019.02.313

[3] EU Waste Framework directive. 2025. European Commission Luxembourg. Available from: Waste Framework Directive - Environment - European Commission [Accessed: September 26, 2025]

[4] Economist. Global Waste Generation Will Nearly Double by 2050. London: Economist; 2018. Available from: https://www.economist.com/graphic-detail/2018/10/02/global-waste

[5] Kaza S, Yao LC, Bhada-Tata P, Van Woerden F. What a Waste 2.0: A Global Snapshot of Solid Waste Management to 2050. NW Washington: World Bank; 2018. DOI: 10.1596/978-1-4648-1329

[6] Giusti L. A review of waste management practices and their impact on human health. Waste Management. 2009;**29**(8):2227-2239. DOI: 10.1016/j.wasman.2009.03.028, ISSN 0956-053X

[7] Zhang Z, Malik MZ, Khan A, Ali N, Malik S, Bilal M. Environmental impacts of hazardous waste, and management strategies to reconcile circular economy and eco-sustainability. Science of the Total Environment. 2022;**807**:1-17

[8] HSE. Waste Management and Recycling. Buxton Derbyshire UK. Available from: http://www.hse.gov.uk/waste/: HSE; 2013 [Accessed: July 1, 2025]

[9] EU. Basel Convention on the Control of Transboundary Movements of Hazardous Wastes and their Disposal. Luxembourg: EU; Publications Office of the European Union; 1993. Available from: www.eur-lex.europa.eu/EN/legal-content/summary/basel-convention-on-the-controlof-transboundary-movements-ofhazardous-wastes-and-their-disposal.Html

[10] EUROSTAT. Waste Generation by Economic Activities and Households, EU, 2022 (% Share of Total Waste). Png - Statistics Explained. Luxembourg: EUROSTAT; 2025. Available from: www.ec.europa.eu/eurostat/statistics-explained/index.php?title=File:waste_generation_by_economic_activities_and_household,_EU,_2022_(%25_share_of_total_waste).png [Accessed: July 23, 2025]

[11] Edjabou ME, Boldrin A, Astrup TF. Compositional analysis of seasonal variation in Danish residual household waste. Resources, Conservation and Recycling. 2018;**130**(2018):70-79. DOI: 10.1016/j.resconrec.2017.11.013, ISSN 0921-3449

[12] Inglezakis VJ, Moustakas K. Household hazardous waste management: A review. Journal of Environmental Management. 2015;**150**(2015):310-321. DOI: 10.1016/j.jenvman.2014.11.021, ISSN 0301-4797. Available from: https://www.sciencedirect.com/science/article/pii/S0301479714005623

[13] Jawjit S, Narom N, Thongkaow P. Evaluating household hazardous

waste generation, composition, and health risks in an urban municipality. Journal of Human, Earth, and Future. 2024;**5**(3):471-482. DOI: 10.28991/HEF-2024-05-03-011

[14] Zionmarketresearch. Hazardous Waste Management Market Size, Trends and Forecast 2034. 2025. Available from: www.zionmarketresearch.com/report/hazardous-waste-management-market [Accessed: July 23, 2025].

[15] Wald Schlager K, Lechthaler S, Stauch G, Schüttrumpf H. The way of microplastic through the environment – Application of the source-pathway-receptor model (review). Science of the Total Environment. 2020;**713**(2020):136584. DOI: 10.1016/j.scitotenv.2020.136584, ISSN 0048-9697. Available from: https://www.sciencedirect.com/science/article/pii/S0048969720300942

[16] UN(EMP). An Overview of UN Activities and Initiatives Related to Pollution. 2023. Available from: https://unemg.org/wp-content/uploads/2023/05/Mapping-report-'An-Overview-of-UN-Activities-and-Initiatives-related-to-Pollution-v3.pdf [Accessed: August 12, 2025]

[17] OECD. Management of Pharmaceutical Household Waste: Limiting Environmental Impacts of Unused or Expired Medicine. Paris FRANCE: OECD; 2022

[18] Yu Z, Gao C, Yang C, Zhang L. Insights into quantities, flows and recycling technology of E waste in China for resource sustainable society. Journal of Cleaner Production. 2023;**393**:136222

[19] Kelly TM, Winchester JW. Can hazardous waste sites be breached as a result of future climate change? Journal of Environmental Engineering. 2005;**131**(5). DOI: 10.1061/(ASCE)0733-9372(2005)131:5(810)

[20] Neksumi M, Zishan M, Sushmita B, Manzoor U. The global menace of hazardous waste: Challenges and management. In: Baskar C, Ramakrishna S, Baskar S, Sharma R, Chinnappan A, Sehrawat R, editors. Handbook of Solid Waste Management. Singapore: Springer; 2022. DOI: 10.1007/978-981-16-4230-2_112

[21] Gichamo T, Gökçekuş H. Interrelation between climate change and solid waste. Journal of Environmental Pollution and Control. 2019;**2**(1):104

[22] Hussain CM, Paulraj MS, Nuzhat S. Source Reduction and Waste Management. Holland: Elsevier; 2021

[23] EU (2024). Ecodesign for sustainable products regulation background and aims online Regulation (EU) 2024/1781 of the European Parliament and of the Council of 13 June 2024 establishing a framework for the setting of ecodesign requirements for sustainable products, amending Directive (EU) 2020/1828 [Accessed: August 12, 2025].

[24] INDAVOR. Municipal Waste Management. Cheshire UK: INDAVOR Middlewich; 2025 . Available from: https://indaver.com/services/municipalwaste-management [Accessed: August 14, 2025]

[25] UN. Sustainable Development Goals. United Nations Development Programme. 2025. Available from: www.undp.org/sustainable-development-goals [Accessed: July 23, 2025]

[26] IEA. Energy Security: Reliable, Affordable Access to all Fuels and Energy Sources. Paris, France: International Energy Agency; 2022. Available from: https://www.iea.org/topics/

energy-security [Accessed: August 12, 2025]

[27] CEWP. 2022. Available from: https://www.cewep.eu/municipal-waste-treatment-2022/ [Accessed: August 15, 2025]

[28] Mendes Campolina J, São Leandro Sigrist C, de Faulstich Paiva JM, et al. A study on the environmental aspects of WEEE plastic recycling in a Brazilian company. International Journal of Life Cycle Assessment. 2017;**22**:1957-1968. DOI: 10.1007/s11367-017-1282-2

[29] Lakhouit A. Enhancing environmental benefits through recycling glass to improve soil properties and concrete. Cleaner Waste Systems. 2025;**11**:100301. DOI: 10.1016/j.clwas.2025.100301. ISSN 2772-9125

[30] Ayilara MS, Olanrewaju OS, Babaloa OO, Odeyemi O. Waste management through composting: Challenges and potentials. Sustainability. 2020;**12**(11):4456

[31] ECN (European Compost Network). Plastics, Microplastics in Compost and Digestate. ECN Position Paper. 2021. Available from: https://www.compostnetwork.info/download/ecn-position-paper-plasticsmicroplastics-in-compost-and-digestate/ [Accessed August 13, 2025]

[32] EPA. Landfill Manuals Landfill Site Designs. 2025. Available from: https://www.epa.ie/publications/compliance--enforcement/waste/landfill-manuals-landfill-site-design.php [Accessed: July 9, 2025]

[33] Treacy. Wastewater Operation Requirements and Distribution in Urban Water Reuse Handbook. 1st ed. Boca Raton Florida. Available from: https://www.taylorfrancis.com/chapters/edit/10.1201/b19646-47/wastewater-operation-requirements-distribution-systems-josephine-treacy?context=ubx&refId=5c1a1618-1cef-438a-8bc6-577f7f1b823d: CRC Press; 2016 [Accessed: July 23, 2025]

[34] APEC Water. Solvents Poured down the Household Drain May Cause Pollution. California United States: APEC Water; 2025

[35] Regional Waste Management Planning Offices Ireland. Household Hazard Waste. 2025. Available from: https://mywaste.ie/what-to-do-with-different-types-of-waste/item/household-hazardous-waste/ [Accessed: August 13, 2025]

[36] Mmereki D, Baldwin A, Hong L, et al. The management of hazardous waste in developing countries. Management of Hazardous Wastes. InTech; 2016. DOI: 10.5772/63055

[37] Energy Ireland. Transforming Waste into Energy: Dublin Waste-to-Energy's Vision for the Future. Energy Ireland: Lower Fitzwilliam st. Dublin Ireland

[38] Chen D, Yin L, Wang H, He P. Pyrolysis technologies for municipal solid waste: A review. Waste Management. 2014;**34**(12):2466-2486. DOI: 10.1016/j.wasman.2014.08.004. ISSN 0956-053X

[39] Arcus. Feasibility Study on a Piolet Plant. Hindenburgstr. Ludwigsburg: ARCUS Greencycling Technologies; 2025

[40] Al-Rumaihi A, Shahbaz M, Mckay G, Mackey H, Al-Ansari T. A review of pyrolysis technologies and feedstock: A blending approach for plastic and biomass towards optimum biochar yield. Renewable and Sustainable Energy Reviews. 2022;**167**:112715. DOI: 10.1016/j.rser.2022.112715. ISSN 1364-0321.

[41] Situmorang YA, Zhao Z, Chaihad N, Wang C, Anniwaer A, Kasai Y, et al. Steam gasification of co-pyrolysis chars from various types of biomass. International Journal of Hydrogen Energy. 2021;**46**(5):3640-3650. DOI: 10.1016/j.ijhydene.2020.10.199. ISSN 0360-3199

[42] Hameed Z, Aslam M, Khan Z, Maqsood K, Atabani AE, Ghauri M, et al. Gasification of municipal solid waste blends with biomass for energy production and resources recovery: Current status, hybrid technologies and innovative prospects. Renewable and Sustainable Energy Reviews. 2021;**136**:110375. DOI: 10.1016/J.RSER.2020.110375

[43] NAP. Waste Incineration and Public Health. Washington DC: The National Academies Press; 2025

[44] Wajda A, Brociek R, Pleszczyński M. Optimization of energy recovery from hazardous waste in a waste incineration plant with the use of an application. PRO. 2022;**10**:462. DOI: 10.3390/PR10030462

[45] Chen Z, Tong K, Xu F, Xue M, Chen H, Chen Q, et al. Development of supercritical water oxidation technology for application to hazardous waste treatment: An extreme case study. Journal of Environmental Chemical Engineering. 2021;**9**(4):105296. DOI: 10.1016/j.jece.2021.105296. ISSN 2213-3437

[46] Li Y, Duan Y, Wang S, Zhang F, Li J, Dai Z, et al. Supercritical water oxidation for the treatment and utilization of organic wastes: Factor effects, reaction enhancement, and novel process. Environmental Research. 2024;**251**(Pt 1):118571. DOI: 10.1016/j.envres.2024.118571. Epub 2024 Mar 1

[47] World Economic Forum. What Is Supercritical Water Oxidation and How Can it Ensure a Truly Circular Economy. Geneva Switzerland. Available from: www.weforum.org/stories/2023/07supercritical-wateroxidation-ensure-a-truly-circular-economy/: World Economic Forum; 2023 [Accessed: July 21, 2025]

[48] Svanström M, Fröling M, Olofsson M, Lundin M. Environmental assessment of supercritical water oxidation and other sewage sludge handling options. Waste Management & Research. 2005;**23**(4):356-366. DOI: 10.1177/0734242X05054324

[49] Benmakhlouf N, Outili N, García-Jarana B, Sánchez-Oneto J, Portela JR, Jeguirim M, et al. Applications of supercritical water in waste treatment and valorization: A review. Energies. 2023;**16**(4):2081. DOI: 10.3390/en16042081

[50] Farmer JC. Electrochemical treatment of mixed and hazard waste. Studies in Environmental Science. 1994;**59**:565-600

[51] Saurabh Yadav S, Suantak Kamsonlian S. A review of electrochemical methods for treatment of wastewater. Materials Today: Proceedings. 2023;**78**(Part 1):36-39. DOI: 10.1016/j.matpr.2022.11.036. ISSN 2214-7853. Available from: https://www.sciencedirect.com/science/article/pii/

[52] SEAI. Carbon Capture Utilization and Storage (CCUS). Dublin Ireland: SEAI; Hatch Street; 2025

[53] López-Portillo MP, Martínez-Jiménez G, Ropero-Moriones E, Saavedra-Serrano MC. Waste treatments in the European Union: A comparative analysis across its member states. Heliyon. 2021;**7**(12):e08645. DOI: 10.1016/j.heliyon.2021.e08645, ISSN 2405-8440. Available from: https://

www.sciencedirect.com/science/article/pii/S2405844021027481

[54] Britannica. Environmental Law. Definition, History, Principles, Examples, & Facts. Britannica; 2025

[55] Esmaeilian B, Wang B, Lewis K, Duarte F, Ratti C, Behdad S. The future of waste management in smart and sustainable cities: A review and concept paper. Waste Management. 2018;**81**:177-195. DOI: 10.1016/j.wasman.2018.09.047, ISSN 0956-053X

[56] EPA. Waste Classification List of Waste & Determining if Waste Is Hazardous or Non-hazardous. County Wexford, Ireland: EPA; Johnstown Castle Estate; 2015. Available from: https://www.epa.ie/pubs/reports/waste/stats/wasteclassification/EPA_Waste_Classification_2015_Web.pdf

[57] Britannica. Hazardous-Waste Management - Treatment, Storage, Disposal. Britannica; 2025

[58] EU. Packaging and Packaging Waste Directive 94/62/EC (PPWD) the Packaging and Packaging Waste Regulation 2025/40 (PPWR), Packaging Waste. Chicago USA: European Commission; 2025

[59] Agovino M, Cerciello M, Musella G, Garofalo A. European waste management regulations and the transition towards circular economy. A shift-and-share analysis. Journal of Environmental Management. 2024;**354**:120423. DOI: 10.1016/j.jenvman.2024.120423

[60] EUR-Lex Hazardous Waste. EUR-Lex; Office of the European Union. Luxembourg; 2025

[61] HAS. (Ireland) REACH. Health and Safety Authority; James Joyce Street. Dublin Ireland; 2025

[62] OECD. Extended Producer Responsibility and Economic Instruments OECD Policy Paper No. 41. Paris FRANCE: OECD; 2024

[63] Basel. Basel Convention on the Control of Transboundary Movements of Hazardous Wastes and their Disposal. Switzerland: Basel Convention Offices; 2025

[64] UNEP. Stockholm Convention on Persistent Organic Pollutants (POPS); Texts and Annexes Revised in 2023. Switzerland: United Nations Environmental Programme (UNEP); 2023

[65] Halkos GE, Aslanidis P-SC. Reviewing the integrated institutional waste-related framework for circular economy in the European Union. Waste Management Bulletin. 2024;**2**(3):28-35. DOI: 10.1016/j.wmb.2024.06.002, ISSN 2949-7507

[66] Treacy J. Policy implementation on waste management and achievement of related SDGs. In: Leal Filho W, Dinis MAP, Moggi S, Price E, Hope A, editors. SDGs in the European Region. Implementing the UN Sustainable Development Goals – Regional Perspectives. Cham: Springer; 2023. DOI: 10.1007/978-3-030-91261-1_35-1

[67] Plastic Europe. How Pyrolysis Oil Is Produced from Mixed Plastic Waste. 2025. Available from: www.plasticseurope.org/de/how-pyrolysis-oil-is-produced-from-mixed-plastic-waste/ [Accessed: July 21, 2025]

[68] UNEP. Global Waste Management Outlook 2024. Kenya: UNEP-UN Environment Programme; 2024. Available from: www.unep.org/resources/global-waste-management-outlook-2024 [Accessed: September 26, 2025]

[69] HAS: EPA. Guidance for the Management of Household Hazardous Waste at Civic Amenity Sites. 2020. Dublin Ireland. Available from: https://www.epa.ie/publications/monitoring--assessment/waste/hazardous-waste/HSA_EPA_Hazardous_Waste_Guidance_CAS.pdf [Accessed: July 23, 2025]

[70] Gholampour Arbastan H, Gitipour S. Evaluating the consequences of household hazardous waste diversion on public health and ecological risks of leachate exposure. International Journal of Environmental Science and Technology. 2022;**19**:4407-4420. DOI: 10.1007/s13762-022-04063-5

[71] Alam P, Ahmede K. Impact of solid waste on health and the environment. International Journal of Sustainable Development and Green Economy. 2013;**2**:165-168

[72] Sultana S, Islam MS, Jahan F, Khatun F. Awareness and practice on household solid waste management among the community people. The Open Nursing Journal. 2021;**11**:349-366. DOI: 10.4236/ojn.2021.115031

[73] Lazíková J, Rumanovská LU. Waste in the EU law. Juridical Tribune-Review of Comparative and International Law. 2024;**14**(2):260-285

[74] EC. Circular Economy Action Plan. European Commission: 2025 Publications Office of the European Union Luxembourg

[75] Bennett GF, G.F. Health aspects of the disposal of waste chemicals. Journal of Hazardous Materials. 1986;**17**(3):340-341. DOI: 10.1016/0304-3894(88)85014-3. ISSN 0304-3894

[76] Das S, Lee S-H, Kumar PK, Soo K-H, Satya S, Bhattacharya L, et al. Solid waste management: Scope and the challenge of sustainability. Journal of Cleaner Production. 2019;**228**:658-678. DOI: 10.1016/j.jclepro.2019.04.323, ISSN 0959-6526

[77] Chew KW, Chia SR, Chia WY, Cheah WY, Munawaroh HSH, Ong WJ. Abatement of hazardous materials and biomass waste via pyrolysis and co-pyrolysis for environmental sustainability and circular economy. Environ Pollution. 2021;**278**:116836. DOI: 10.1016/J.ENVPOL.2021.116836

[78] Martinho GA, Pires AL, Saraiva LR, Ribeiro R. Composition of plastics from waste electrical and electronic equipment (WEEE) by direct sampling. Waste Management. 2012;**32**(6):1213-1217. DOI: 10.1016/j.wasman.2012.02.010, ISSN 0956-053X

## Further readings

[1] HAS Guidance for the Management of Household Hazardous Waste at Civic Amenity Sites - Health and Safety Authority online. Available from: www.hsa.ie/eng/publications_and_forms/publications/chemical_and_hazardous_substance/guidance_for_the_management_of_household_hazardous_waste_at_civic_amenity_sites.html [Accessed: October 17, 2025]

[2] CSIS. Centre for Strategic and International Studies. The Basel Convention: From Hazardous Waste to Plastic Pollution. CSIS; 2021. p. 2021. Available from: www.csis.org/analysis/basel-convention-hazardous-waste-plastic-pollution#:~:text=The%20Basel%20Convention%20unifies%20the,transboundary%20movement%20of%20toxic%20materials [Accessed: October 17, 2025]

[3] Lamma OA. Waste disposal and landfill: Potential hazards and their impact on groundwater. International Journal of Geography, Geology and Environment. 2021;**3**(2):133-141. Available from: www.geojournal.net/archives/2021.v3.i2.B.68 [Accessed: October 17, 2025]

[4] Triantafyllidis KS, Samaras P, Georgios P, Zouboulis A. Application of the pyrolysis technology for industrial and hazardous waste. In: 6th International Conference on Sustainable Solid Waste Management at Naxos Island. Greese; 2018. Available from: www.researchgate.net/publication/326225740_Application_of_the_pyrolysis_technology_for_industrial_and_hazardous_waste_management [Accessed: October 17, 2025]

[5] United Nations Environment Programme. Basel Convention on the Control of Transboundary Movements of Hazardous Wastes and their Disposal: Protocol on Liability and Compensation for Damage Resulting from Transboundary Movements of Hazardous Wastes and their Disposal. United Nations Environment Programme; 1992. Available from: https://www.basel.int/Portals/4/Basel%20Convention/docs/text/BaselConventionText-e.pdf

[6] Fawole AA, Orikpete OF, Ehiobu NN, Ewim DRE. Climate change implications of electronic waste: Strategies for sustainable management. Bulletin of the National Research Centre. 2023;**2023**(47):147. DOI: 10.1186/s42269-023-01124-8. Available from: https://bnrc.springeropen.com/articles/10.1186/s42269-023-01124-8 [Accessed: October 17, 2025]

[7] Zhang J, Chen C, Zhang X, Liu S. Study on the environmental risk assessment of lead-acid batteries. Procedia Environmental Sciences. 2016;**31**:873-879. Available from: www.sciencedirect.com/science/article/pii/S1878029616001043 [Accessed: October 17, 2025]

[8] Sharolyn M, Ahmad K, Mahmood G, Trivedi RC. Municipal solid waste management in Indian cities - a review. Waste Management. 2008;**28**(2):459-467. DOI: 10.1016/j.wasman.2007.02.008. Available from: www.sciencedirect.com/science/article/abs/pii/S0956053X07000645 [Accessed: October 17, 2025]

Chapter 2

# Efficient Management of the Public Municipal Solid Waste Management Service in Cameroon: Concrete Actions to Be Taken

*Joel Sotamenou*

## Abstract

This chapter proposes concrete actions that can promote efficient management of the public municipal solid waste management service in Cameroon, central Africa. The financial, institutional, physical, and technical constraints of the municipal solid waste management service in Yaoundé are presented. The 20 proposed actions are based on a performance indicator, which is based on four modules (institutional, social, economic, and environmental) and 14 objectives. These include, among other things, improving access to solid waste management services for as many households as possible, ensuring proper spatial distribution of waste bins, effectively implementing sanctions against all polluters, improving the level of investment in waste management infrastructure, institutionalizing pre-collection and composting with financial participation from households, and respecting the principles of sustainable development.

**Keywords:** solid waste management, public service, households, pre-collection, composting, Cameroon

## 1. Introduction

In many African countries, the provision of certain public services, such as municipal solid waste (MSW) management, is delegated to private sector companies. In recent years, we have witnessed the development of an organizational model for MSW management that leads to participatory governance involving the municipality, authorized collection companies, pre-collection structures, and households; and leads to a renewal of the public-private partnership (3P) [1–3]. According to the World Bank's 2017 report, the 3Ps improve access to sustainable and quality infrastructure services, all of which contribute to the well-being of citizens. At the heart of the 3Ps is the contract or a public service delegation (PSD) agreement. In theory, the PSD can take many forms: concession, leasing, and public management.

Among these different forms of PSD, concession and leasing are the most common contracts implemented in Africa for the management of MSW. The concession mandates a private company to provide the public MSW collection service under the control of the municipality, which has the right to oversee the management of the service. The company itself finances, at its own risk, the creation of the network, which does not yet exist, and ensures its management. It is remunerated by a price paid by the service users, and generally benefits from a monopoly. Unlike the concession, in the leasing contract, the private company that performs the public service benefits from the facilities necessary for the operation of the service and receives a fixed fee in return.

Despite the convincing results in MSW management obtained in certain countries such as Rwanda, South Africa or Ghana [2], it is still necessary to question the social performance of such an institutional arrangement in a context where interests diverge, since the State seeks to maximize the well-being of its citizens while private companies only seek to maximize their profits. It is therefore up to each state or municipality to put in place coordination and motivation mechanisms in order to reduce the distortions that can arise from this type of relationship. Tirole [4] suggests that solutions to these opportunistic behaviors include ensuring that detailed contracts are written, including indexation clauses on inflation or on the cost of supply, and establishing clear procedures for review and recourse to arbitration. Obviously, the renewal of this type of contract can only be done after a multi-criteria evaluation. The evaluation in question involves the definition of objective functions or performance criteria and the selection of performance indicators.

According to Boulanger [5], an indicator is a measurable variable used to account for a phenomenon or an action. In this sense, it fulfills a basic information function for political decision-making, as well as an evaluation or audit function, internal and/or external. It can also constitute a set of elements of the collective definition of a public service (goals to be achieved, standards to be respected) and the means of achieving it (measurement of well-being).

Sotamenou [6] proposes a tool for evaluating the public MSW management service based on a set of criteria (objectives to be achieved) and indicators (action variables), which seem essential to the establishment of an effective and sustainable MSW management service in the African context.

## 2. The major obstacles to the municipal solid waste management service in Yaoundé, Cameroon

The MSW management in Yaoundé suffers from a number of financial, institutional, physical, and technical constraints, which are important to note.

The main limitations to MSW financing in Yaoundé include the low level of the Household Waste Collection Tax (HWCT) withheld at source for public and private sector employees, and for establishments paying business licenses. The fact that the HWCT does not cover the majority of households, and especially the informal sector, is a weakness. There is also the problem of having a single fund within the municipalities themselves, and between the municipal funds and the public treasury. The annual budget for MSW management in Yaoundé, for example, is 15 billion FCFA (25 million USD) for a daily collection of approximately 1200 tons (60% of the daily MSW production). Since formal composting and recycling activities are nonexistent, it becomes difficult for the government, which finances 85% of this budget (only 15% for municipalities) to regularly mobilize the funds necessary for optimal collection of MSW.

At the institutional level, many actors are involved in the MSW management sector, with roles that are not clearly defined. At the state level, almost all ministerial departments are involved in MSW management, although to varying degrees. At the level of decentralized local authorities, the overlapping responsibilities of the urban community and the urban district municipalities' give rise to conflicts between the appointed government delegate and the elected mayors of the urban district municipalities. However, it is important to mention the lack of clear boundaries for the exercise of responsibilities by the various stakeholders through specific laws and texts. The noted imprecisions are a source of misunderstanding, power struggles, and underperformance in the public management of MSW. The lack of coordination of the actions of the various stakeholders and the multiplicity of decision-making centers are a source of evasion of responsibility or a struggle for competence.

Physical obstacles to MSW management in Yaoundé include, among others: the rugged terrain consisting of a series of hills and valleys with altitudes varying between 700 and 1200 m, the inadequacy of service roads, which open up certain neighborhoods and deprive them of MSW disposal services, and the continuous expansion of the urban perimeter, which increases MSW collection distances. MSW collection methods and means must therefore be rethought to take these physical constraints into account. According to Sotamenou [3], in a context where unstructured housing in Yaoundé houses more than 60% of the capital's population, and more than half of the inhabitants access their homes *via* a neighborhood track, pre-collection (collection and disposal of MSW from isolated areas to a transfer center, a garbage bin or an area accessible to collection trucks) is unfortunately not institutionalized. Taking such activity into account involves the construction in the lowlands of waste transfer or grouping centers, which are spaces designed to receive and process (sorting and composting) pre-collected waste.

It is noted that MSW collection services in Yaoundé suffer from a lack of mastery and inadequacy of appropriate techniques. According to [7], MSW collection equipment must be adapted to the nature of the waste produced. Technical obstacles to public MSW management in Yaoundé include, among others: the rapid increase in the population, which in turn increases the quantity of MSW to be collected, the spatial distribution of garbage bins, which often does not take into account the needs of users (quantity of waste produced per person or per household, distance to travel to find a garbage bin, accessibility to collection trucks, incivility of the population). Collection tools are very often inadequate for the nature of the MSW to be collected.

Following the constraints related to MSW management in Yaoundé, let us now present the 20 major actions that must be implemented for effective MSW management in Yaoundé in particular and in Cameroon in general. But first, let us present the methodology adopted.

## 3. The methods

While many studies propose performance indicators for the public drinking water supply service [8, 9], very few similar studies exist in the field of waste. However, performance evaluation methods in the waste sector have been proposed, notably in Italy [10], Australia [11], the United States [12], or France [13]. The approach followed in this chapter is inspired by the article by Bertolini [13], which proposes a simple system of indicators for the evaluation of the performance of a public service for local household waste management in France. The tool he proposes is a dashboard made up of four components (economic/financial, user comfort, social, and environmental) and ten performance criteria and indicators. At the end of his study, the author indicates that the proposed method is not necessarily adapted to the specific context of cities in developing countries. Based on the tool he proposes and the specificities of African cities, Sotamenou [6] identifies concrete actions to be implemented for efficient management of the public MSW management service in Africa in general and in Cameroon in particular.

## 4. The 20 concrete actions that can promote efficient management of the public municipal solid waste management service in Cameroon

These 20 concrete actions are included in modules, each consisting of objectives to be achieved.

### 4.1 The institutional module

Three objectives have been identified for the institutional module: regulation and selection of partners, cost transparency, and state accountability. The actions to be implemented to achieve these institutional objectives are:

#### *4.1.1 Regulation and selection of partners*

*Actions 1 and 2: Regularly launching calls for tenders and periodically evaluating specifications.*

A transparent and reasoned selection of service providers is crucial. As such, calls for tenders must transparently enable the selection of the most relevant offer and the awarding of associated contracts. In the event that the public sector engages the services of one or more private companies, as is the case in Cameroon, the parties involved must favor long-term agreements, given the scale of the required investments, with regular assessments of compliance with the specifications and revisions when necessary.

#### *4.1.2 Cost transparency*

*Actions 3 and 4: Clarity of the contract and incompleteness of the contract.*

Transparency is necessary for the accountability of MSW public service management. A relevant estimate of the costs of a waste management action requires not only that the State or local authority know the investment and operating costs of the unit operations included in the action in question, but also that the specifications relating to the action (e.g., on the quantities of waste to be collected and treated, and the geographical areas concerned). Where possible, it is also important to consider the externalities generated by the treatment systems in question (landfill, composting, etc.) and to anticipate the hazards that may arise during the period. But the principle of transparency goes beyond the realm of costs. Indeed, local authorities and citizens are entitled to receive precise, up-to-date information on the performance of the MSW management system with regard to environmental protection and human health.

#### *4.1.3 The responsibility of the State and/or decentralized local authorities (DLAs)*

*Actions 5 and 6: Strong involvement of the State and/or DLAs; and development of State and/or DLAs subsidies.*

Given the characteristics of the MSW management field, particularly with regard to its impacts on health and social well-being, and its environmental issues, the government has a responsibility toward citizens to guarantee the protection of their quality of life, and this over the long term. To this end, it must clarify in advance the role of all actors involved in MSW management, because the generally noted imprecisions are a source of misunderstanding, power struggles, and underperformance. The lack of coordination of the actions of the different stakeholders, the multiplicity of decision-making centers, is also a source of responsibility leakage, or competency struggles. Since MSW production is positively correlated with urbanization, it requires a more significant and progressive involvement of the State and/or the DLAs.

### 4.2 The social module

The social module is based on the principle of equity, which includes the following objectives: universal access to a MSW collection service, user participation in decision-making, the polluter pays principle, and intergenerational equity. The actions to be implemented to achieve these social objectives are:

#### *4.2.1 Universal access to an MSW collection service*

*Actions 7 and 8: The percentage of the population served, and the number and distribution of garbage bins.*

The greatest number of households must have permanent access to the waste collection service. In Cameroon, for this to be effective, it is necessary to take into account the physical constraints (existence of hills and lowlands, low percentage of drivable roads, etc.) and technical constraints (collection equipment sometimes unsuitable for physical constraints, etc.) related to the MSW management. Indeed, the spatial distribution of garbage bins does not often take into account the needs of users (quantities of waste produced per person or per household, accessibility to collection trucks, incivility of populations, etc.). The distribution density of garbage bins is also an important criterion since the distance separating a household from a garbage bin significantly influences the delivery of MSW to the collection point.

#### *4.2.2 User participation in decision-making*

*Actions 9 and 10: The presence of local elected officials and associations on the board of directors of the collection company(ies); and the size of the communications budget of these company(ies).*

An MSW management service must take into account the wishes of the population when implementing any collection strategy. In this sense, the presence of elected officials and representatives of associations on the management or monitoring committee of the collection company is necessary, because they would act as a transmission belt between the company and the population. The frequency of communication and awareness-raising operations, environmental education, and even public health campaigns also makes it possible to measure the degree of commitment of the collection company to social issues.

### *4.2.3 The polluter pays principle*

*Action 11: Effective implementation of sanctions (fines) against any polluter.*

This principle is generally accepted as a fundamental principle of fairness. Poor MSW management can have harmful consequences not only from an economic point of view but also from a health point of view. In order to address the pollution problem, the polluter pays principle is very important. According to this principle, the user pays for the waste they produce. Everyone is therefore encouraged to control their own production and reduce the nuisances linked to their activity; it is therefore a fairly reliable incentive mechanism. This principle can help to overcome the shortcomings linked to the financing of MSW management by involving households, as in many West African cities, where their contributions sometimes cover up to 50% of the overall cost of MSW management. For example, the population of Yaoundé, which includes one million households (the National Institute of Statistics estimates the average size of a household in Yaoundé at nearly five people; while "World Population Review" estimates the current population of Yaoundé at nearly 5 million), a monthly contribution per household that varies from 500 FCFA to 1000 FCFA (between one and 20 USD), depending on the type of housing, would make it possible to collect between 500 million FCFA (one million USD) and one billion FCFA (two million USD) every month. Considering an average monthly contribution of 750 FCFA (1.5 dollars), this represents 750 million FCFA (1.5 million USD) per month, or nearly nine billion FCFA per year (18 million USD), which represents nearly 60% of the current public budget dedicated to MSW management in Yaoundé. Revenues from the sale of compost (more than 80% of MSW is made up of organic matter) and recycled waste will also contribute to increasing the budget of this alternative financing system. The polluter pays principle could also consist of finding a fine mechanism to be paid by any citizen who disposes of their waste in an anarchic manner.

### *4.2.4 Intergenerational equity (access for all generations, present and future, to a quality waste management service)*

*Action 12: The level of investment in MSW management infrastructure.*

This concerns the level of investment in the construction and development of landfills and collection centers, the renewal of collection equipment, etc. In the Cameroonian context, where more than half of the daily MSW remains in the neighborhoods, the development and construction in the lowlands of transfer or collection centers for waste which are spaces designed to receive and treat (sorting and composting) pre-collected waste. It is undeniable that future generations will be affected by the current MSW management. Moreover, when investments in MSW management infrastructure such as garbage bins and transfer centers are very often neglected and postponed to the future, the risk of having to face a major crisis, or having to repair at great expense the consequences that result from such negligence, is increased. Similarly, the increasing evolution of the quantities of MSW produced implies a regular renewal of collection equipment.

## 4.3 The economic module

The economic module includes the following objectives to be achieved: the implementation of pre-collection and composting, allocative efficiency, and statistical efficiency. The actions to be implemented to achieve these economic objectives are:

#### *4.3.1 The implementation of pre-collection*

*Action 13: The degree of support for pre-collection (remuneration of pre-collectors, purchase of pre-collection equipment, medical care for pre-collectors, income from the sale of recovered objects, etc.).*

Given the importance of pre-collection in the MSW management systems in Cameroon, it is urgent to institutionalize it. Pre-collection associations must be recognized as being of public utility. Indeed, the relevance of pre-collection in a context marked by the physical and technical constraints mentioned above allows many households to make a living from the recovery and recycling of pre-collected plastic, glass, and aluminum objects. Pre-collection activities also provide employment for young people in the neighborhoods and thus improve their living standards. Due to its importance, the implementation of pre-collection is therefore a major element to be taken into account in the public MSW management service in Cameroon [14].

#### *4.3.2 Setting up composting*

*Action 14: the degree of implementation of composting (remuneration of composters, purchase of composting equipment, medical care of composters, income from the sale of compost, etc.).*

Just as important as pre-collection, composting, which is the transformation of the fermentable portion of pre-collected MSW into compost, must be institutionalized. It will also be necessary to ensure that the compost produced is of good quality. The relevance of composting in African economies in general and in Cameroon in particular is justified by the very high proportion of the fermentable fraction contained in MSW (more than 80%). MSW transfer centers in lowlands would facilitate composting operations and would make quality compost available to farmers at a lower cost (since the proximity of the compost production site to agricultural areas such as the lowlands means that transport costs, for example, are significantly reduced). Compost also presents significant economic challenges [15, 16]. As with pre-collection, the lack of composting operations, both centralized (at the landfill) and decentralized (in transfer centers), is a major handicap to the economic performance of the public MSW management service in Cameroon.

#### *4.3.3 Allocative efficiency*

*Action 15: Minimization and control of management costs (pre-collection, collection, transport, and routing of final waste to the landfill).*

In economic terms, a method is said to be efficient if, and only if, there is no production technique that allows it to generate the maximum output (high collection rates) using fewer inputs (fuel, maintenance of collection trucks, labor, etc.). While efficiency is said to be static when there is no waste of resources or the minimization of waste management costs, allocative efficiency, however, implies more than the minimization of costs; it also requires that the logistics related to the collection of MSW be greater in the most unsanitary areas.

#### *4.3.4 Statistical efficiency*

*Action 16: Collection rate above 75% and increasing over time.*

The MSW collection rate is the ratio between the quantities collected and produced. For [17], this rate must be greater than 75% because below this threshold, the accumulated MSW becomes harmful to the environment and the health of the population. To be able to manage MSW satisfactorily, the level of collection must evolve with the rate of production. The government can, therefore, for example, index the waste collection rate] to the rate of urbanization.

## 4.4 The environmental module

The environmental module includes the following objectives: eco-efficiency, reduction of greenhouse gas emissions, and reduction of underground pollution. The actions to be implemented to achieve these environmental objectives are:

### *4.4.1 Eco-efficiency*

*Action 17: Respect for the principles of sustainable development.*

Given the characteristics of the "waste product," efficiency must also be seen from the perspective of sustainable development; hence, the concept of eco-efficiency, which is based on the idea of respecting and preserving the environment. It is undeniable that MSW production is increasing and therefore represents a major challenge to the quality of our environment. In order to avoid a global crisis, it is essential to adopt a sustainable waste management strategy as soon as possible. Sustainable MSW management refers to MSW management that meets the needs of the present without compromising the ability of future generations to meet their own needs. The concept of eco-efficiency implies the integration of the environment and the economy, the protection of people's health, the protection of ecosystems, compliance with international standards and the promotion of equity.

### *4.4.2 Reducing greenhouse gas emissions*

*Actions 18 and 19: The low rate of waste burial and the implementation of the biogas treatment system.*

An effective MSW management system must integrate the objective of reducing greenhouse gas emissions. As mentioned above, MSW in Cameroon consists mainly of organic matter. Burying this waste causes toxic gas emissions such as methane, which contribute to global warming. Any effective MSW management system must therefore ensure upstream awareness among populations to reduce their MSW production and downstream implementation of biogas recovery systems in landfills.

### *4.4.3 Reducing underground pollution*

*Action 20: Implementation of a leachate treatment system.*

Leachate is the residual liquid generated by the percolation of water and liquids through a waste storage area, chemical products or simply soil contaminated by pollutants. This product of the dissolution of organic matter and trace elements (heavy metals, organic and chemical pollutants, radionuclides, etc.) is a source of soil and water pollution, including groundwater. Given the high level of water contained in MSW in Cameroon (more than 50% of the total mass), they produce enormous quantities of leachate that infiltrate into the soil and pollute groundwater. It is therefore essential to set up a reliable system for the recovery and treatment (decontamination) of this leachate in landfills or collection centers.

## 5. Conclusion

This chapter revisits the paper of Sotamenou [6] and proposes concrete actions that can promote effective public municipal solid waste management service in Cameroon. It first highlights the financial, institutional, physical, and technical obstacles before suggesting 20 actions. These include, among others, improving access to the solid waste management service for the greatest number of households, ensuring good spatial distribution of garbage bins, and the effective implementation of sanctions against all polluters, increasing the level of investment in MSW management infrastructure, and especially the institutionalization of pre-collection and composting, with the financial participation of households.

## Author details

Joel Sotamenou
University of Yaoundé II, Soa, Cameroon

*Address all correspondence to: sotamenou@yahoo.fr

## References


[1] Bah RD. Du partenariat public-privé a la gouvernance participative: cas du secteur des déchets ménagers dans la ville d'Abidjan [thesis]. Paris, France: Université de Versailles; 2009

[2] Sotamenou J. La gestion des déchets solides en Afrique cinquante ans après les indépendances: bilan et perspectives. Afrique durable. 2017;**2030**(1):21-37

[3] Sotamenou J. Le compostage: une alternative soutenable de gestion publique des déchets solides au Cameroun [thesis]. France: Université de Yaoundé II (Cameroun) et CIRAD; 2012, 286 p

[4] Tirole J. Concession de services publics: quel contrat l'Etat doit-il passer pour garantir un service optimal? Problemes Economiques. 1999;**2**(460):10-16

[5] Boulanger PM. Les indicateurs de développement durable: un défi scientifique, un enjeu démocratique. Belgique: Les séminaires de l'Iddri, n° 12 Institut pour un développement durable; 2004

[6] Sotamenou J. Proposition d'un outil d'évaluation du service publique de gestion des déchets solides municipaux en Afrique. Déchets Sciences et Techniques. 2018;**76**:33-42

[7] Cointreau-Levine S, de Kadt M. Living with Garbage: Cities learn to recycle. Development Forum. January–February 1991:12-13

[8] Guerin-Schneider L, Nakhla M. Les indicateurs de performance: une evolution clef dans la gestion et la regulation des services d'eau et d'assainissement. Flux. 2003;**52**:55-68

[9] Mathieu-Makkaoui R. Reflexion sur les formes d'action collective et la gouvernance des ressources en eau dans les pays en developpement [thesis]. Versailles, France: Universite Versailles Saint-Quentin-en-Yvelines; 2009, 283 p

[10] Chifari R, Renner A, Piano SL, Ripa M, Bukkens SG, Giampietro M. Development of a municipal solid waste management decision support tool for Naples, Italy. Journal of Cleaner Production. 2017;**161**:1032-1043

[11] Zaman AU. Identification of key assessment indicators of the zero waste management systems. Ecological Indicators. 2014;**36**:682-693

[12] Greene KL, Tonjes DJ. Quantitative assessments of municipal waste management systems: Using different indicators to compare and rank programs in New York state. Waste Management. 2014;**34**(4):825-836

[13] Bertolini G. Service public local des déchets ménagers. A la recherche d'indicateurs de performance. Développement durable et territoires. [En ligne], Points de vue. 2008. (2003-2010):1-16. DOI: 10.4000/developpementdurable.5743

[14] Sotamenou J, Molua LE, Akamin A. Application of analytic hierarchy process decision model for solid waste management strategy in Yaoundé, Cameroon. The Journal of Solid Waste Technology and Management. 2019;**45**(4):502-517

[15] Sotamenou J, De Jaeger S, Rousseau S. Drivers of legal and illegal solid waste disposal in the global south-the case of households in Yaoundé (Cameroon). Journal of Environmental Management. 2019;**240**:321-330

[16] Thuries L, Ganry F, Sotamenou J, Olivier R, Parrot L, Simon S, et al. Cash for trash: An agro-economic value assessment of urban organic materials used as fertilizers in Cameroon. Agronomy for Sustainable Development. 2019;**39**(6):52

[17] Hebette A. Guide pratique de la gestion des déchets solides urbains en Afrique Subsaharienne. Lux-Development. 1996. 149p

Section 2

# Sustainable Energy Solutions and Integrated Waste Valorization

Chapter 3

# Harnessing Nonfood Biomass for Biodiesel Production: Integration of Hazardous Waste Treatment into Sustainable Energy Solutions

*Maria Ameen, Mushtaq Ahmad, Muhammad Zafar, Muhammad Faisal Amin, Muhammad Saad Amin, Shabir Ahmad, Mamoona Munir, Muhammad Iqbal, Huma Qureshi, Nabila Mahmood, Faten Zubair Filimban, Salman Majeed, Wajia Noor and Bibi Sadia*

## Abstract

This chapter explores the integration of hazardous waste treatment into biodiesel production using nonfood biomass as a sustainable feedstock. Nonfood biomass—such as agricultural residues, lignocellulosic waste, forestry byproducts, and microalgae—represents a promising alternative to food-based biofuels, enabling energy production without compromising food security. Concurrently, the incorporation of hazardous waste streams, including industrial effluents, contaminated biomass, and organic toxins, into bioenergy pathways offers an innovative approach to environmental remediation. Through advanced pretreatment, bioremediation, and catalytic transesterification processes, these waste materials can be converted into high-quality biodiesel. This integrated framework not only addresses the challenges of waste management but also enhances energy circularity, reduces greenhouse gas emissions, and contributes to carbon neutrality. The chapter highlights recent technological advancements, system efficiencies, and sustainability metrics in waste-to-energy systems, providing a roadmap for implementing closed-loop bioenergy strategies. By aligning with circular economy principles and global climate goals, this approach offers a dual benefit—clean energy generation and effective hazardous waste mitigation.

**Keywords:** biodiesel production, nonfood biomass, hazardous waste treatment, sustainable energy, waste-to-energy

## 1. Introduction

The escalating global demand for sustainable and low-carbon energy solutions has positioned biodiesel as a pivotal alternative to fossil fuels. Among the various strategies to produce biodiesel, the utilization of nonfood biomass is gaining significant attention due to its dual advantages: minimizing the food-versus-fuel conflict and contributing to environmental sustainability [1]. Feedstocks such as agricultural residues, forestry waste, algae, and organic industrial byproducts offer abundant, renewable, and underutilized resources for biodiesel synthesis [2].

Simultaneously, the integration of hazardous waste treatment into biodiesel production processes presents a transformative approach to managing environmental pollution. Hazardous wastes—including contaminated biomass, industrial effluents, and toxic organic compounds—can be processed through biochemical and thermochemical pathways, not only reducing their ecological impact but also recovering valuable energy [3, 4]. This synergy between biofuel production and waste remediation exemplifies the principles of a circular economy, where waste streams are repurposed into functional energy carriers (**Figure 1**).

Recent developments in catalyst design, transesterification technology, and biorefinery integration have enhanced the feasibility and efficiency of converting complex waste materials into biodiesel. Policy frameworks and international climate commitments are further driving investment and innovation in waste-to-energy systems. By aligning renewable energy generation with waste mitigation strategies, this integrated approach supports carbon neutrality, energy resilience, and long-term ecological balance [5].

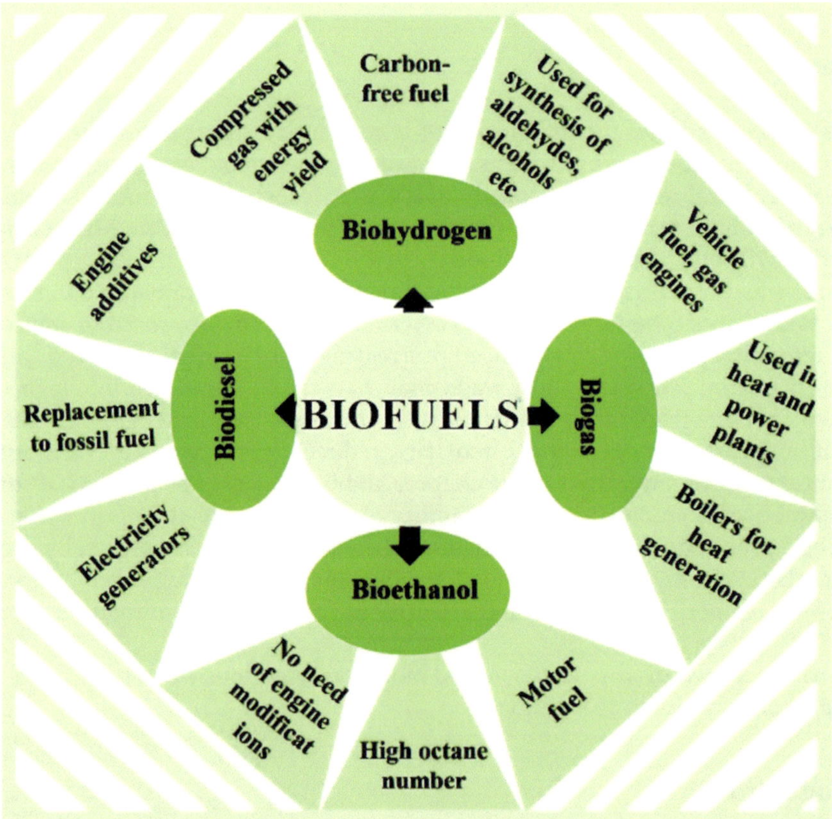


**Figure 1.**
*Schematic flowsheet for various applications of biofuels.*

This chapter explores the scientific basis, technological advancements, and environmental implications of using nonfood biomass for biodiesel production, with a particular focus on incorporating hazardous waste treatment into the process. Through this lens, it aims to highlight sustainable bioenergy pathways that address both energy needs and waste management challenges.

## 2. Nonfood biomass as a sustainable feedstock for biodiesel

Nonfood biomass represents a critical and sustainable alternative to traditional food-based feedstocks in biodiesel production [6]. Unlike crops such as soybean, palm, or rapeseed oil—which compete with food systems and arable land—nonfood biomass utilizes resources that are otherwise considered waste or low-value byproducts. These include agricultural residues like rice husks and wheat straw, lignocellulosic waste such as corn stover, bagasse, and sugarcane tops, as well as forestry residues like sawdust and bark. High-oil-content nonedible plant species such as *Jatropha curcas*, *Pongamia pinnata*, *Calophyllum inophyllum* (Alexandrian laurel), and *Ricinus communis* (castor bean) offer promising oil yields without affecting food supplies [7]. Additionally, energy crops like *Miscanthus giganteus*, *Sorghum bicolor*, *Agave tequilana*, and *Camelina sativa* are gaining attention for their adaptability to marginal lands and low input requirements (**Table 1**).

| Sr. no. | Biomass source | Oil yield (L/ha/year) | Common usage |
|---|---|---|---|
| 1 | *Jatropha curcas* | 1500–1900 | Widely used for biodiesel; drought-tolerant crop |
| 2 | *Pongamia pinnata* | 1200–1500 | Biodiesel, soil reclamation, intercropping |
| 3 | *Ricinus communis* (Castor) | 1200–1800 | Biodiesel, lubricants, pharmaceuticals |
| 4 | *Calophyllum inophyllum* | 1000–1500 | Biodiesel, traditional medicine, lamp oil |
| 5 | *Camelina sativa* | 900–1500 | Biodiesel, aviation biofuel |
| 6 | *Azadirachta indica* (Neem) | 400–800 | Biodiesel, pesticides, medicinal uses |
| 7 | *Madhuca indica* (Mahua) | 1200–1600 | Biodiesel, cooking oil (nonedible), rural energy |
| 8 | *Simmondsia chinensis* (Jojoba) | 800–1200 | Biodiesel, cosmetics, lubricants |
| 9 | *Chlorella vulgaris* (Algae) | 30,000–60,000 | High-yield biodiesel, wastewater treatment |
| 10 | *Nannochloropsis sp.* (Algae) | 20,000–40,000 | Biodiesel, omega-3 extraction, carbon capture |
| 11 | *Corn Stover* | <100 (via pyrolysis/oil blend) | Blended biodiesel, bio-oil, cellulosic ethanol |
| 12 | *Sugarcane Bagasse* | <50 (indirect oil via gasification) | Biofuel blend, cogeneration |
| 13 | *Miscanthus giganteus* | <200 (cellulosic feedstock) | Cellulosic biofuel, combustion |
| 14 | *Agave tequilana* | <150 (sugar-rich biomass) | Ethanol, biogas, bio-crude oil |
| 15 | *Sorghum bicolor* (Sweet Sorghum) | 400–600 | Biodiesel, ethanol, fodder |

**Table 1.**
*Nonfood biomass sources for biodiesel production [7–10].*

Microalgae such as *Chlorella vulgaris* and *Nannochloropsis* are particularly noteworthy, capable of yielding 20 to 30 times more oil per hectare than traditional oilseed crops while growing in saline or wastewater. Globally, agricultural activities generate over 4 billion tons of residues annually, much of which remains underutilized or openly burned, contributing to air pollution. Lignocellulosic biomass alone accounts for approximately 70–80% of all plant biomass on Earth, offering an enormous, renewable, and carbon-rich source of energy [11].

From an environmental standpoint, using nonfood biomass helps minimize greenhouse gas emissions significantly. Life cycle assessments have demonstrated that biodiesel derived from lignocellulosic and nonedible oil sources can reduce emissions by up to 86% compared to fossil diesel. Additionally, the cultivation and processing of nonfood biomass support carbon neutrality, as the $CO_2$ absorbed during plant growth offsets the emissions released during combustion. Economically, oil-producing species such as *Jatropha*, *Pongamia*, and *Calophyllum* can produce between 1200 and 1900 liters of oil per hectare annually, with significantly lower input requirements than food crops [7, 12]. Furthermore, integrating these feedstocks into existing agricultural systems can lower feedstock costs by 30–50%, create rural employment, and add value to agricultural and forestry waste streams [13]. Thus, nonfood biomass not only advances the sustainability of biodiesel production but also contributes to environmental conservation, rural development, and the global shift toward a circular bioeconomy.

## 3. Types and sources of hazardous waste suitable for bioenergy conversion

The integration of hazardous waste into biodiesel production represents a transformative shift in how we perceive waste—not merely as a disposal problem, but as a valuable resource for sustainable energy. Hazardous waste, typically associated with health and environmental risks, can be repurposed through innovative technologies into feedstocks for bioenergy, including biodiesel. These waste types originate from diverse sources such as industrial operations, agricultural runoff, healthcare facilities, and municipal systems, and many contain significant amounts of organic material or combustible components amenable to bioenergy recovery.

Industrial effluents and sludge from chemical, pharmaceutical, textile, and petrochemical industries are among the most abundant hazardous waste streams with bioenergy potential. For example, the textile industry produces large volumes of dye-laden sludge rich in hydrocarbons and fatty acids, which can be converted into biodiesel precursors after detoxification [14]. Similarly, waste oils and greases from food processing, leather treatment, or metalworking industries—classified as hazardous due to contamination—can be processed via transesterification after pretreatment. Globally, millions of tons of waste oil are generated annually, with a significant portion still improperly disposed of or incinerated [15].

Another promising category is contaminated biomass—organic materials exposed to pesticides, heavy metals, or pathogens during their lifecycle. For instance, spent crops from agrochemical-intensive farming or spoiled food waste from municipal sources often become toxic due to fungal or bacterial contamination. Through bioremediation and thermal pretreatment, these materials can be rendered safe and transformed into useful biofuels [16]. Agricultural practices themselves generate large quantities of animal manure and pesticide-laden crop residues, both of which can be subjected to anaerobic digestion or pyrolysis, leading to the generation of biogas or bio-crude that can be upgraded into biodiesel [17].

Medical and laboratory waste, particularly lipid-rich biological discards and expired pharmaceuticals, is another underutilized stream. Although challenging due to regulatory constraints and contamination risks, with proper sterilization and fractionation technologies, selected fractions can contribute to lipid-based biofuel production. Municipal solid waste (MSW), especially the organic fraction, contributes more than 1.3 billion tons annually worldwide. Of this, a significant portion is comprised of organic hazardous waste—including paints, solvents, batteries, and expired food contaminated with chemicals—which can be separated and valorized into energy carriers.

Incorporating these hazardous waste streams into biodiesel systems not only diverts them from landfills and incinerators but also reduces greenhouse gas emissions and mitigates environmental contamination [17]. The key lies in advanced pretreatment techniques—such as detoxification, membrane filtration, and catalytic cracking—that make hazardous inputs viable for energy conversion [18].

## 4. Technological pathways: From biomass and waste to biodiesel

The successful conversion of hazardous biomass into biodiesel hinges on effective pretreatment and detoxification processes, essential for improving feedstock quality and ensuring a safe, high-yield biofuel [19]. Hazardous biomass—such as industrial sludge, pesticide-laden residues, and contaminated food waste—often contains heavy metals, pathogens, complex hydrocarbons, or synthetic chemicals that need to be removed or neutralized to enable efficient biodiesel production.

*Physical separation:* One of the first methods in pretreatment is physical separation, which involves techniques such as sedimentation, centrifugation, and filtration. For example, in the treatment of oily industrial sludge, centrifugation can separate up to 95% of free oil from the waste stream, leaving behind a solid waste that is much easier to process. In some cases, size reduction (shredding) may be used to increase the surface area of solid biomass, improving the efficiency of subsequent chemical or biological treatments [20].

*Chemical pretreatment:* In many cases, biomass contaminated with complex organic chemicals requires chemical pretreatment to release free fatty acids or neutralize toxic compounds. One widely used method is alkaline hydrolysis, which can increase the yield of biodiesel from waste oils by 30–40% by breaking down triglycerides into free fatty acids [21]. For example, the textile industry generates waste sludge containing synthetic dyes, which can be neutralized by adjusting the pH using alkaline treatments, thus reducing toxicity and improving biodiesel yield. In electroplating wastes, heavy metals such as chromium and cadmium can be removed by applying chelation agents, which can bind to these metals with an efficiency of over 90% [22, 23].

*Biological pretreatment:* For more complex hazardous biomasses, biological treatments can be highly effective. White-rot fungi, like Phanerochaete chrysosporium, are known to degrade up to 80% of lignin and phenolic compounds present in contaminated biomass, reducing toxicity and improving biodiesel yield. *Pseudomonas putida*, a bacterium capable of breaking down hydrocarbons, is frequently used to bioremediate petroleum-contaminated residues. Bioremediation techniques, when used alongside chemical and physical treatments, can reduce the toxicity of waste materials by 50–70%, making them more suitable for biodiesel production [24].

*Thermal methods:* Pyrolysis and steam explosion are thermal processes used for detoxifying biomass, particularly when dealing with highly contaminated organic waste. Pyrolysis, for example, can destroy pathogens and complex organics and is capable of reducing hazardous biomass to biochar—which can be further processed into bio-oil for conversion into biodiesel [25]. The thermal degradation of medical waste, such as expired pharmaceuticals or lipid-rich biological discards, typically results in the removal of 95% of harmful contaminants. In the case of pesticide-laden agricultural waste, steam explosion breaks down cell walls and facilitates oil extraction, increasing oil yield by up to 40% compared to untreated feedstocks.

*Advanced pretreatment technologies:* Newer methods, such as microwave-assisted extraction or ultrasonic treatment, have demonstrated high efficiency in extracting oils from hazardous waste. In one study, microwave-assisted extraction of oil from waste oils and food residues increased biodiesel yield by 25–35% when compared to conventional methods. Hydrothermal liquefaction has also shown promise in converting wet waste, such as algae or food residues, into bio-oil with efficiencies reaching 85–90% oil recovery from total biomass [26].

Hence, the choice of pretreatment depends on the nature of the hazardous waste being processed, with each technique offering specific advantages in terms of efficiency, environmental safety, and cost. The use of these pretreatment and detoxification methods helps mitigate the risks associated with hazardous biomass while ensuring high-quality biodiesel production [27]. Additionally, successful detoxification reduces the environmental impact, such as heavy metal contamination, by up to 90%, enabling the conversion of potentially harmful waste into a renewable energy source.

## 5. Pretreatment and bioremediation techniques for contaminated biomass

Catalytic transesterification remains the cornerstone of biodiesel production, involving the reaction of triglycerides with short-chain alcohols (typically methanol or ethanol) in the presence of a catalyst to form fatty acid methyl esters (FAME) and glycerol. The process efficiency is strongly influenced by feedstock quality, alcohol-to-oil molar ratio, catalyst type, reaction temperature, and moisture content. For conventional base-catalyzed transesterification using refined oils, optimal conditions typically include a 6:1 methanol-to-oil ratio, 1% w/w NaOH or KOH catalyst, 60–65°C reaction temperature, and 1-hour residence time, yielding conversion rates as high as 98–99%. However, hazardous biomass feedstocks often contain high levels of free fatty acids (FFAs), water, and contaminants, which deactivate base catalysts and induce saponification, drastically reducing biodiesel yield and quality [28].

Although pretreatment significantly reduces toxicity, trace contaminants such as heavy metals (e.g., Pb, Cr, Cd) may persist in the final biodiesel if not properly managed. These can pose risks to engine performance and emissions control systems. Therefore, strict fuel quality testing and adherence to EN 14214/ASTM D6751 standards are essential, and further studies are required to assess long-term ecological and mechanical impacts of waste-derived biodiesel.

To overcome these limitations, heterogeneous and acid-catalyzed systems have been increasingly adopted. Solid acid catalysts like sulfonated carbon, heteropoly acids, or metal oxides (e.g., ZnO, TiO, AlO) offer resilience against FFAs and water while allowing for easy separation and catalyst reuse. For instance, sulfonated biochar catalysts have demonstrated FAME yields of 92–95% from high-FFA feedstocks (>15%) under mild conditions (65°C, 4–6 hours). Similarly, bifunctional catalysts combining acid and base sites enable simultaneous esterification of FFAs and transesterification of triglycerides, effectively converting low-quality oils from hazardous biomass. Enzymatic catalysis using lipases (e.g., *Candida antarctica* lipase B) provides high specificity and minimal byproduct formation but is currently constrained by high costs, long reaction times (>12 h), and enzyme deactivation in the presence of heavy metals and toxins—commonly found in hazardous waste oils [29]. Immobilized enzymes, which allow repeated use and enhance tolerance to inhibitors. Microbial engineering, especially genetically modified strains of Pseudomonas or Bacillus, to improve hydrocarbon degradation and toxin resistance during bioremediation. Despite the challenges associated with enzymatic catalysis, recent research has focused on enhancing enzyme resilience through immobilization techniques and the use of protective supports such as mesoporous silica, biochar, and magnetic nanoparticles. These supports not only stabilize the enzyme structure under harsh processing conditions but also facilitate recovery and reuse, improving the overall economics of enzymatic biodiesel production. Moreover, advances in protein engineering and synthetic biology have enabled the development of mutant lipases with improved activity against high-FFA and toxin-rich substrates. For instance, engineered variants of *Candida antarctica* lipase B have shown increased tolerance to metal ions and organic solvents, making them more suitable for hazardous biomass conversion. Coupled with microreactor systems and continuous flow biocatalysis, these innovations promise to reduce reaction times and scale enzymatic processes for industrial application, offering a greener and more selective alternative to traditional chemical catalysis.

Emerging catalytic technologies extend beyond conventional FAME production, exploring non-transesterification pathways like hydrodeoxygenation (HDO) and catalytic pyrolysis, which enable the production of renewable diesel and green hydrocarbons. For instance, HDO using catalysts such as $NiMo/Al_2O_3$ or $CoMo/\gamma\text{-}Al_2O_3$ under high-pressure hydrogen (4–6 MPa) and temperatures of 350–400°C effectively removes oxygen from bio-oils derived from pyrolyzed hazardous biomass, producing hydrocarbon fuels that are fully compatible with existing diesel engines and infrastructure. Studies report hydrocarbon yields of 80–85% from pyrolyzed algal oil and plastic-contaminated biomass using these catalysts. Moreover, plasma-assisted catalysis and nanostructured catalysts are under development to enhance reaction kinetics, increase surface area activity, and reduce catalyst fouling in systems dealing with impure or mixed hazardous waste streams. These advanced catalytic strategies represent a critical step in closing the loop between waste detoxification and high-efficiency fuel production, pushing the boundaries of biodiesel chemistry toward next-generation, waste-integrated biorefineries.

## 6. Catalytic transesterification using mixed waste inputs

Efficient biodiesel production from hazardous and nonfood biomass requires robust process integration to minimize energy consumption, reduce emissions, and enhance yield. A typical biodiesel production system encompasses multiple interdependent units—pretreatment, transesterification, separation, purification, and waste recovery—each of which can introduce inefficiencies if not properly synchronized [30]. Using process integration techniques such as pinch analysis, plants can reduce energy demand by 20–30% through optimized heat exchange between hot and cold process streams. Similarly, mass integration allows the reuse of process water and unreacted methanol, achieving up to 90% solvent recovery, which is critical when dealing with high-moisture hazardous feedstocks where water removal is energy-intensive [31].

Advanced process simulation tools such as Aspen Plus, CHEMCAD, and SuperPro Designer are now widely employed to model and optimize biodiesel plants, particularly those handling heterogeneous waste inputs. Dynamic modeling enables real-time prediction of system behavior under variable feedstock conditions, crucial when working with hazardous waste streams where composition can fluctuate significantly. For instance, colocating waste oil collection, pretreatment, and biodiesel conversion units into a single continuous process reduces transportation losses and improves lifecycle energy efficiency by up to 35%. Furthermore, integrating glycerol purification (the main byproduct) and converting it into value-added chemicals like propanediol or epichlorohydrin enhances economic viability and contributes to circularity in system design [32].

Optimization is further supported by integrating control systems and smart automation, including real-time spectroscopic monitoring (e.g., FTIR, NIR) and PID/PLC control loops, which enable continuous monitoring of conversion efficiency, catalyst performance, and contaminant levels. Studies on digitally controlled pilot plants have reported yield improvements of 12–18% over traditional batch systems, alongside a 15–20% reduction in process downtime due to automated fault detection [33]. Energy and exergy analyses of such systems indicate that overall process efficiency can exceed 78%, even when low-quality or hazardo us feedstocks are used. Ultimately, integrated biodiesel production systems must balance thermodynamic efficiency, emissions control, economic feasibility, and regulatory compliance to ensure scalability and sustainability in real-world deployment.

Recent pilot-scale systems have demonstrated the feasibility of converting industrial oily sludge and food-contaminated biomass into biodiesel. For instance:

i. A 5-ton/day pilot facility in South Asia colocated with a textile industry utilized bifunctional catalysts and achieved over 85% biodiesel yield from dye-laden sludge after alkaline and biological pretreatment.

ii. In Europe, a modular mobile transesterification unit deployed near agro-industrial zones processed 1000 liters/day of pesticide-contaminated oilseed waste, successfully integrating detoxification, microwave-assisted oil extraction, and enzymatic transesterification.

These pilot projects underscore the scalability potential of decentralized, waste-based biodiesel units, particularly when colocated with waste sources to minimize logistics costs and environmental risks.

## 7. Environmental and energy performance metrics

Environmental performance assessment of waste-derived biodiesel is essential to justify its role in sustainable energy systems. Life Cycle Assessment (LCA), standardized under ISO 14040/44 protocols, evaluates the environmental impacts of the entire biodiesel production chain—from biomass acquisition and pretreatment to fuel combustion [34]. When using hazardous and nonfood waste as feedstocks, LCA studies consistently show lower global warming potential (GWP) compared to fossil diesel. For instance, biodiesel from waste cooking oil demonstrates a 70–90% reduction in $CO_2$-equivalent emissions per MJ of fuel produced, while systems using industrial oily sludge or contaminated lignocellulosic biomass show GWP savings of 55–80%, depending on detoxification intensity and energy input [35].

Beyond GWP, LCA also accounts for eutrophication potential (EP), acidification potential (AP), human toxicity (HTP), and fossil energy demand (FED). Waste-based biodiesel systems significantly outperform food-based or virgin oil pathways in these categories. For example, in a comparative LCA between soybean biodiesel and biodiesel from waste tallow, the latter showed 86% lower eutrophication impacts and 94% less land use. Acidification potential is also reduced by minimizing the need for fertilizers and pesticides, particularly when using contaminated agricultural residues [36]. However, these benefits can be offset if the pretreatment and transesterification stages are not energy-optimized—highlighting the importance of coupling LCA with process modeling and sensitivity analysis to identify environmental hotspots.

Crucially, LCA of hazardous biomass-to-biodiesel pathways must account for co-benefits related to waste mitigation. When hazardous wastes are diverted from incineration or landfilling to biofuel production, the avoided emissions and leachate toxicity contribute to the system's net environmental benefit. Moreover, integrated systems that valorize byproducts such as glycerol, biochar, or biogas further improve environmental performance metrics. These findings position waste-based biodiesel as not only a low-carbon alternative but also a functional tool for waste management and pollution reduction, especially in circular bioeconomy frameworks [37].

While detoxification processes such as pyrolysis and chemical neutralization require significant energy inputs, these are often balanced by integrated heat recovery systems, solvent recycling, and the avoidance of emissions from landfilling or incineration. Life cycle modeling indicates that optimized detoxification contributes only 10–20% to total greenhouse gases (GHG) emissions of the system, and coproduct valorization (e.g., biochar use in soil carbon sequestration) can further reduce net emissions.

## 8. Circular economy integration and carbon neutrality goals

The integration of hazardous biomass-based biodiesel into the circular economy (CE) framework is grounded in the principle of valorizing waste as a resource while minimizing environmental externalities. Unlike linear models, where waste is treated as a terminal burden, CE enables continuous material flows—transforming industrial effluents, agricultural toxins, and organic sludges into feedstocks for renewable fuel production. In such systems, mass and energy flows are internally looped, ensuring that residues (e.g., spent biomass, glycerol, biochar) are either reprocessed, upcycled, or used for cogeneration. Empirical case studies indicate that closed-loop biodiesel plants incorporating glycerol-to-biochemicals pathways and biochar reuse for soil remediation can reduce lifecycle emissions by up to 80%, while simultaneously diverting 5–10 tons of hazardous waste per day from incineration or landfill [38].

Biodiesel derived from hazardous or nonfood biomass plays a pivotal role in national carbon neutrality strategies, as it displaces fossil diesel in heavy transport, agriculture, and off-grid sectors [39]. Carbon accounting based on intergovernmental panel on climate change (IPCC) Tier 2 methodologies shows that per MJ of energy, waste-based biodiesel emits just 15–25 g$CO_2$e, compared to 94 g$CO_2$e for conventional diesel—a 75–85% reduction. When the avoided emissions from hazardous waste treatment are included (e.g., methane avoidance from landfilling, incineration offsets), some systems achieve net-negative GHG balances [36]. For example, converting pesticide-laden rice straw to biodiesel and using the biochar residue for carbon sequestration in soil can yield net carbon savings of 0.6–0.9 t $CO_2$e per ton of feedstock. This dual benefit aligns directly with the Paris Agreement and net-zero targets being adopted by over 140 countries [40].

Key enablers of circularity include digital twin frameworks, material flow analysis (MFA), and industrial symbiosis, where waste streams from one sector (e.g., textile effluent sludge) become inputs for another (e.g., biodiesel and nutrient recovery) [41]. Lifecycle circularity indices (LCI) and resource recovery efficiency (RRE) metrics are increasingly used to quantify performance—studies report LCI values of 0.72–0.85 for advanced biodiesel systems, indicating that over 70% of input materials are recirculated. Integration with urban waste management infrastructures, co-digestion facilities, or agro-industrial clusters can scale these models while enhancing resilience [42]. Ultimately, embedding hazardous biomass valorization into circular economy platforms not only decarbonizes transport fuels but also provides a strategic route to holistic waste management and systemic sustainability [43].

## 9. Challenges, opportunities, and future directions

### 9.1 Challenges

#### *9.1.1 Feedstock variability and standardization*

Hazardous biomass sources such as industrial sludges, pesticide-contaminated residues, and wastewater scum vary widely in chemical composition and toxicity, complicating standardization and requiring advanced characterization and detoxification.

#### *9.1.2 Catalyst deactivation due to contaminants*

High levels of impurities (e.g., heavy metals, sulfur, phosphorus) can poison catalysts used in transesterification or pyrolysis, reducing efficiency and increasing the frequency of catalyst regeneration or replacement.

#### *9.1.3 High energy and chemical inputs for pretreatment*

Many hazardous waste feedstocks require multistage pretreatment (e.g., dewatering, pH neutralization, toxin degradation), which increases the energy and chemical input costs, potentially offsetting environmental benefits [44].

#### *9.1.4 Regulatory barriers and permitting complexity*

Ambiguous or restrictive legislation around hazardous waste classification often leads to long permitting processes, legal uncertainty, and underinvestment in pilot-scale facilities or commercial rollout.

#### *9.1.5 Public perception and social acceptance*

The term "hazardous waste" carries stigma, which can hinder public and stakeholder support, particularly when biodiesel is intended for public transport or municipal applications.

#### *9.1.6 Scale-up and operational complexity*

While lab-scale processes show promise, the transition to commercial scale introduces complexity in feedstock logistics, system integration, continuous operation, and risk management for hazardous materials. Moreover, operational risks such as chemical exposure, flammability, and biohazard contamination require strict adherence to occupational safety measures, including the use of personal protective equipment (PPE), automated closed-loop systems, and secure storage under controlled temperature and pressure. Worker training and emergency protocols must be part of standard operating procedures, particularly when handling toxic sludges or pharmaceutical residues.

#### *9.1.7 High capital and operating costs of waste-based systems*

Building integrated waste-to-biodiesel plants with containment, emission controls, and specialized equipment requires significantly higher capital expenditures (CAPEX) compared to conventional biodiesel plants.

While comprehensive plant-specific cost data vary with region, scale, and feedstock type, general estimates suggest:

CAPEX for integrated hazardous-waste biodiesel plants ranges from USD 1500–2500 per ton/year of installed biodiesel capacity—higher than conventional systems due to additional detoxification and safety infrastructure.

OPEX is highly feedstock-dependent but can be significantly reduced through the use of negative-cost inputs (e.g., industrial waste oils) and coproduct valorization (e.g., glycerol → epichlorohydrin, biochar → soil amendment). In optimized systems, production costs can range from USD 0.60–0.85 per liter, competitive with fossil diesel (USD 0.80–1.20/liter, depending on subsidies and crude oil prices).

### 9.2 Opportunities

#### *9.2.1 Low-cost feedstocks availability*

Hazardous biomass is often available at negative or zero cost due to waste disposal fees paid by generators, significantly lowering the raw material expenditure for biodiesel production.

#### *9.2.2 Dual environmental benefits*

These systems offer a dual benefit—energy generation and waste detoxification—which aligns with sustainable development goals (SDGs) and can attract climate financing and circular economy investments.

#### *9.2.3 Carbon credit potential*

Waste-based biodiesel projects can earn carbon credits under CDM, VERRA, or national offset schemes, particularly when GHG reductions and waste abatement are both verifiable.

### *9.2.4 Emerging catalytic technologies*

Emerging technologies—such as bifunctional catalysts, ionic liquids, enzymatic transesterification, and microbial pretreatment—can enhance process robustness and tolerance to feedstock impurities.

### *9.2.5 Government funding support*

National and international programs are increasingly allocating funds for waste-to-energy innovations, including tax relief, grants, and subsidies for demonstration plants.

### *9.2.6 Industrial colocation synergy*

Colocation with industries generating hazardous biomass (e.g., textiles, tanneries, agrochemicals) allows integrated waste valorization, reducing logistics costs and maximizing circularity.

### *9.2.7 Byproduct valorization pathways*

Glycerol, biochar, and nutrient-rich digestates from the process can be converted into high-value chemicals, fuels, or soil amendments, improving overall techno-economic viability.

## 9.3 Future directions

- Development of universal protocols and tools (e.g., rapid spectroscopy, toxicity indices) to assess hazardous biomass suitability for biodiesel production will streamline industrial adoption.

- Implementing real-time digital twins and machine learning algorithms can enhance process optimization, predictive maintenance, and operational safety in complex waste-to-fuel systems.

- Small-scale, modular biodiesel units can be deployed near hazardous waste sources, reducing transport risks and supporting decentralized energy generation in rural or industrial zones.

- Integration with Carbon Capture and Storage (CCS); and future systems may incorporate CCS to move toward true net-negative emissions, particularly when co-firing syngas or utilizing $CO_2$ from process off-gas streams.

- Coordinated international frameworks for hazardous waste valorization and renewable fuel certification can accelerate technology transfer, investment, and trade of green fuels.

- Next-generation life cycle tools will need to evaluate impacts not only at the plant level but across industrial ecosystems, incorporating circularity, toxicity, and socioeconomic dimensions.

- Collaborations between governments, research institutions, and industry players will be critical to de-risk projects, scale infrastructure, and foster innovation in hazardous biomass utilization.

## 10. Conclusion

The integration of hazardous biomass into biodiesel production represents a transformative step toward achieving sustainable energy and waste management goals. By leveraging nonfood, contaminated, or industrial waste streams, this approach addresses two of the most pressing global challenges simultaneously: the growing demand for renewable fuels and the urgent need to safely manage hazardous organic waste. Technological advancements in pretreatment, catalytic transesterification, and bioprocessing have made it increasingly viable to convert complex waste matrices into high-quality biodiesel with low lifecycle emissions and net-positive energy returns.

While advancements in pretreatment and catalytic processes have improved feasibility, key research gaps remain. In particular, improving catalyst resilience, optimizing waste logistics, and developing standardized preprocessing and characterization methods are critical for scalability. Economically, the use of low-cost waste feedstocks and carbon credit opportunities supports the competitiveness of this approach. However, its success will rely not only on interdisciplinary innovation and policy support. Furthermore, alignment with circular economy principles enhances resource efficiency by closing material and energy loops while enabling carbon neutrality pathways through carbon offsetting, biochar sequestration, and avoided methane emissions. Strategic policy frameworks and supportive regulatory environments will be essential in de-risking projects and fostering large-scale deployment.

Looking forward, the success of hazardous biomass-to-biodiesel systems will depend on interdisciplinary innovation, regulatory harmonization, and multi-stakeholder collaboration. With continued investment, standardization, and societal acceptance, this integrated model has the potential to redefine waste as a strategic energy asset—ushering in a new paradigm of clean, circular, and climate-resilient bioeconomy.

## Acknowledgements

The author acknowledges the use of AI for language polishing of the manuscript.

## Conflict of interest

The authors declare no conflict of interest.

## Author details

Maria Ameen[1,2]*, Mushtaq Ahmad[1], Muhammad Zafar[1], Muhammad Faisal Amin[3,4], Muhammad Saad Amin[5], Shabir Ahmad[6], Mamoona Munir[7], Muhammad Iqbal[2], Huma Qureshi[2], Nabila Mahmood[2], Faten Zubair Filimban[8], Salman Majeed[9], Wajia Noor[10] and Bibi Sadia[10]

1 Biofuel Laboratory, Department of Plant Sciences, Quaid-i-Azam University, Islamabad, Pakistan

2 Department of Botany, University of Chakwal, Chakwal, Punjab, Pakistan

3 Joint Doctoral School, Silesian University of Technology, Gliwice, Poland

4 Centre of Polymer and Carbon Materials, Polish Academy of Sciences, Zabrze, Poland

5 Faculty of Agricultural Engineering and Technology, PMAS-Arid Agriculture University, Rawalpindi, Pakistan

6 Northwest Institute of Eco-environment and Resources, Chinese Academy of Sciences (CAS), Lanzhou, China

7 Department of Botany, Rawalpindi Women University, Rawalpindi, Punjab, Pakistan

8 Division of Botany, Department of Biological Sciences, Faculty of Sciences, King Abdulaziz University, Jeddah, Saudi Arabia

9 Department of Botany, University of Mianwali, Mianwali, Pakistan

10 Plant Systematics and Biodiversity Lab, Quaid-i-Azam University, Islamabad, Pakistan

*Address all correspondence to: mameen@bs.qau.edu.pk

## References

[1] Ameen M et al. Cleaner biofuel production via process parametric optimization of nonedible feedstock in a membrane reactor using a Titania-based heterogeneous nanocatalyst: An aid to sustainable energy development. Membranes. 2023;**13**(12):889

[2] Bhagia S et al. Critical review of FDM 3D printing of PLA biocomposites filled with biomass resources, characterization, biodegradability, upcycling and opportunities for biorefineries. Applied Materials Today. 2021;**24**:101078

[3] Guleria A et al. Removal of inorganic toxic contaminants from wastewater using sustainable biomass: A review. Science of the Total Environment. 2022;**823**:153689

[4] Aziz A et al. Microscopic techniques for characterization and authentication of oil-yielding seeds. Microscopy Research and Technique. 2022;**85**(3):900-916

[5] Znad H, Awual MR, Martini S. The utilization of algae and seaweed biomass for bioremediation of heavy metal-contaminated wastewater. Molecules. 2022;**27**(4):1275

[6] Ameen M et al. Prospects of catalysis for process sustainability of eco-green biodiesel synthesis via transesterification: A state-of-the-art review. Sustainability. 2022;**14**(12):7032

[7] Dhanya M. Biodiesel production from non-edible oilseeds. In: Algal Biofuel. UK: CRC Press; 2022. pp. 149-181

[8] Ahmad S et al. Performance and emission characteristics of second-generation biodiesel with oxygenated additives. Energies. 2023;**16**(13):5153

[9] Kumar A, Sharma S. Potential non-edible oil resources as biodiesel feedstock: An Indian perspective. Renewable and Sustainable Energy Reviews. 2011;**15**(4):1791-1800

[10] Noor W et al. Taxonomic significance of palyno-morphic markers for the delimitation of some *Brassicaceous taxa* in Balochistan Province (Pakistan). Genetic Resources and Crop Evolution. 2025;**72**(1):587-614

[11] Fatma S et al. Lignocellulosic biomass: A sustainable bioenergy source for the future. Protein and Peptide Letters. 2018;**25**(2):148-163

[12] Ong HC. Techno-Economic Analysis of Biodiesel Production from Palm, Jatropha Curcas and *Calophyllum inophyllum* for Road Transport in Malaysia/Ong Hwai Chyuan. University of Malaya; 2012

[13] Whalen J et al. Sustainable Biofuel Production from Forestry, Agricultural and Waste Biomass Feedstocks. Elsevier; 2017. pp. 281-283

[14] Tumanyisibwe A et al. Anaerobic digestion of dye wastewater and agricultural waste with bio-energy and biochar recovery: A techno-economic and sustainable approach. Water. 2024;**16**(14):2025

[15] Demirbas A. Waste management, waste resource facilities and waste conversion processes. Energy Conversion and Management. 2011;**52**(2):1280-1287

[16] Ramos A, Monteiro E, Rouboa A. Biomass pre-treatment techniques for the production of biofuels using thermal conversion methods–A review.

Energy Conversion and Management. 2022;**270**:116271

[17] Ameen M et al. Sustainable Innovations in Biodiesel Plants: From Production to Environmental Impact, in Biodiesel Plants-Fueling the Sustainable Outlooks. London, UK: IntechOpen; 2024

[18] di Bitonto L et al. Enhancing biodiesel production from urban sewage sludge: A novel industrial configuration and optimization model. Sustainable Energy Technologies and Assessments. 2023;**60**:103567

[19] Ameen M et al. Biodiesel synthesis from *Cucumis melo* var. Agrestis seed oil: Toward non-food biomass biorefineries. Frontiers in Energy Research. 2022;**10**:830845

[20] Gregori Tena A, Javierre Cazador D. Centrifugal Method to Clean Waste Cooking Oil: A Study of the Feasibility of Constructing a Centrifuge to Remove Unwanted Particles from Waste Cooking Oil Using Reused Materials [Thesis]. 2023

[21] Salaheldeen M et al. Current state and perspectives on transesterification of triglycerides for biodiesel production. Catalysts. 2021;**11**(9):1121

[22] Gulcin İ, Alwasel SH. Metal ions, metal chelators and metal chelating assay as antioxidant method. PRO. 2022;**10**(1):132

[23] Yudaev P, Chistyakov E. Chelating extractants for metals. Metals. 2022;**12**(8):1275

[24] Patel AK et al. Organic wastes bioremediation and its changing prospects. Science of the Total Environment. 2022;**824**:153889

[25] Ameen M et al. Biodiesel Plants-Fueling the Sustainable Outlooks: Fueling the Sustainable Outlooks2025. p. 93

[26] Deepika C et al. Hydrothermal liquefaction of wet microalgal biomass for biofuels and platform chemicals: Advances and future prospects. Discover Applied Sciences. 2024;**6**(5):245

[27] Ameen M et al. Hazardous waste management of novel non-edible *Platycladus orientalis* seed oil via recyclable yttria-based phyto-nanocatalyst: A practical approach towards sustainable bioenergy conversion. Sustainable Energy Technologies and Assessments. 2024;**67**:103845

[28] Omar, Saifuddin A. Optimized Biodiesel Production by Using Ultrasound Assisted Transesterification/ Omar Saifuddin Aasimudin. Universiti Malaya; 2021

[29] Sarmah N et al. Recent advances on sources and industrial applications of lipases. Biotechnology Progress. 2018;**34**(1):5-28

[30] Ameen M et al. Conversion of novel non-edible *Bischofia javanica* seed oil into methyl ester via recyclable zirconia-based phyto-nanocatalyst: A circular bioeconomy approach for eco-sustenance. Environmental Technology & Innovation. 2023;**30**:103101

[31] Musonye FS. Modeling, simulation and optimization of internal heat recovery using process integration technique of pinch analysis [Thesis]. JKUAT-IEET. 2021

[32] Ameen M et al. Assessing the bioenergy potential of novel non-edible biomass resources via ultrastructural analysis of seed sculpturing using microscopic imaging visualization. Agronomy. 2023;**13**(3):735

[33] Chukwunweike J, Abubakar I, Anang AN. Enhancing industrial efficiency through automation: Leveraging PLC, Scada, HMI, batch, and DCS systems for downtime mitigation in industrial control projects. International Journal of Research Publication and Reviews. Sep 2024;**5**(8):1466-1478

[34] Ameen M et al. Wild melon: A novel non-edible feedstock for bioenergy. Petroleum Science. 2018;**15**:405-411

[35] Prasad S et al. Review on biofuel production: Sustainable development scenario, environment, and climate change perspectives– A sustainable approach. Journal of Environmental Chemical Engineering. 2024;**12**(2):111996

[36] Ameen M et al. Oil extraction techniques for bioenergy production: A systematic review. Journal of Biomaterials and Bio-products Technology (JBBT). 2021;**1**(1):1-17

[37] Sikazwe MK, Louw J, Görgens JF. Techno-economic and environmental assessment of a sugarcane biorefinery: Direct and indirect production pathways of biobased adipic acid. Biofuel Research Journal. 2024;**11**(4):2225-2242

[38] Goel S. Solid and hazardous waste management: An introduction. In: Advances in Solid and Hazardous Waste Management. 1st ed. Cham: Springer; 16 Aug 2017. p. 371

[39] Ahmed I et al. Perspective Chapter: Sustainable Management and Utilization of Agricultural Waste–Innovations, Challenges, and Future Prospects. London, UK: IntechOpen; 2025

[40] Dafnomilis I, den Elzen M, van Vuuren D. Paris targets within reach by aligning, broadening and strengthening net-zero pledges. Communications Earth & Environment. 2024;**5**(1):48

[41] Rozina AM, Khan AM, Abbas Q, Arfan M, Mahmood T, Ameen M. Implication of scanning electron microscopy as a tool for identification of novel, nonedible oil seeds for biodiesel production. Microscopy Research and Technique. 2022;**85**(5):1671-1684

[42] Brändström J, Saidani M. Comparison between circularity metrics and LCA: A case study on circular economy strategies. Journal of Cleaner Production. 2022;**371**:133537

[43] Silva S et al. An integrated approach for efficient energy recovery production from livestock and agro-industrial wastes. In: Waste Biomass Management–A Holistic Approach. Springer Nature; 2017. pp. 339-366

[44] Ameen M et al. Prospects of Bioenergy Development in Future. 2023

Chapter 4

# The Use of Non-Food Biomass for Biodiesel Production: Technological and Environmental Challenges [Part 1]

*Asma Nouira, Imen Amri, Imene Bekri-Abbes and Diogo Canavarro*


## Abstract

The urgent global transition towards renewable energy has positioned biodiesel as a pivotal alternative for decarbonizing industrial sector. Moving beyond the contentious "food-versus-fuel" dilemma of first-generation biofuels, this review focuses on the sustainable production of biodiesel from non-food biomass, including agricultural residues, non-edible oilseed crops (e.g., *Jatropha curcas* and *Pongamia pinnata*), waste streams, and microalgae. The chapter provides a systematic analysis of the entire value chain, beginning with a classification of biofuel generations and a critique of food-based feedstocks' socio-economic and environmental limitations. It then details the typology of advanced non-food feedstocks and the technological evolution in conversion processes, highlighting breakthroughs in heterogeneous catalysis and microwave-assisted and ultrasonic transesterification that enhance efficiency and sustainability. The analysis further addresses critical technological and environmental challenges, such as feedstock variability and emission profiles.


**Keywords:** biodiesel, non-food biomass, transesterification, heterogeneous catalysis, circular economy, sustainability

## 1. Introduction

The global energy landscape is undergoing a profound transformation, driven by the imperative to mitigate climate change and enhance energy security. This shift is accelerating the development of sustainable alternatives to fossil fuels, with biodiesel emerging as a critically important solution for decarbonizing the transportation sector [1]. Unlike conventional diesel, biodiesel derived from non-food biomass offers significant environmental advantages, including substantial lifecycle greenhouse gas emission reductions, while strategically avoiding competition with global food supplies [2–8]. The biofuel industry has evolved through distinct generations, each aiming to overcome the limitations of its predecessor. First-generation biofuels, produced from edible crops like corn, sugarcane, and vegetable oils, ignited a widespread debate over the "food-versus-fuel" dilemma, raising concerns over food security, land-use change, and ultimately, their net carbon benefit. This has catalyzed a pivotal re-evaluation, propelling the industry towards more sustainable feedstocks. Second-generation biofuels derived from lignocellulosic biomass and third-generation fuels from algae and photosynthetic microorganisms represent a more sustainable pathway by utilizing non-food resources, agricultural and forestry residues, and waste streams [4]. This chapter comprehensively explores the frontier of non-food biodiesel. It begins by establishing the conceptual framework, defining biofuel generations, and outlining biodiesel's role within the broader global energy transition. It then critically examines the socio-economic and environmental limitations of food-based feedstocks, building a compelling case for the necessity of non-food alternatives. The core of the chapter provides a detailed typology of non-food biomass—from resilient energy crops like Jatropha and Pongamia to waste cooking oils and algal species—and delves into the technological advancements revolutionizing their conversion into biodiesel. This includes a critical analysis of the evolution from homogeneous to heterogeneous catalytic systems and the application of process intensification technologies like microwave and ultrasound. Finally, the chapter addresses the persistent technological and environmental challenges, such as feedstock variability and NOx emissions that must be overcome for widespread commercialization. By synthesizing the current state of knowledge and technology, this review aims to provide a holistic understanding of the opportunities and hurdles in non-food biodiesel production, charting a path for its integration as a sustainable and equitable component of the future energy system.

## 2. Conceptual framework and global issues

### 2.1 Definition and classification of biofuels (first, second, and third generations)

Biofuels are liquid or gaseous fuels derived from biomass, such as agricultural crops including corn, sugarcane, rapeseed, and sunflower, and represent a renewable alternative to fossil fuels [9]. They are typically classified into three generations based on the feedstock and conversion processes involved (**Figure 1**). First-generation biofuels are produced from food crops like corn, sugarcane, and vegetable oils, yielding bioethanol and biodiesel through fermentation and transesterification [10–17]. However, the use of biofuel, instead of minimizing gas emission, it contributes to several environmental and societal problems. A global agricultural model reveals that corn-based ethanol nearly doubles greenhouse gas emissions over 30 years rather than achieving the expected 20% reduction. Similarly, switchgrass biofuels grown on U.S. corn lands raise emissions by 50% due to land-use changes [17]. Furthermore, diverting large portions of crops to fuel production raises concerns over food security and price volatility [13, 14]. Policies are evolving to address these concerns: the European Union has capped first-generation biofuels at 7% of transportation energy consumption to limit land-use impacts. Similarly, the U.S. Renewable Fuel Standard has shifted focus towards advanced biofuels derived from non-food sources, moving away from conventional corn-based ethanol to reduce environmental drawbacks [18–20]. Despite these challenges, first-generation biofuels may maintain relevance

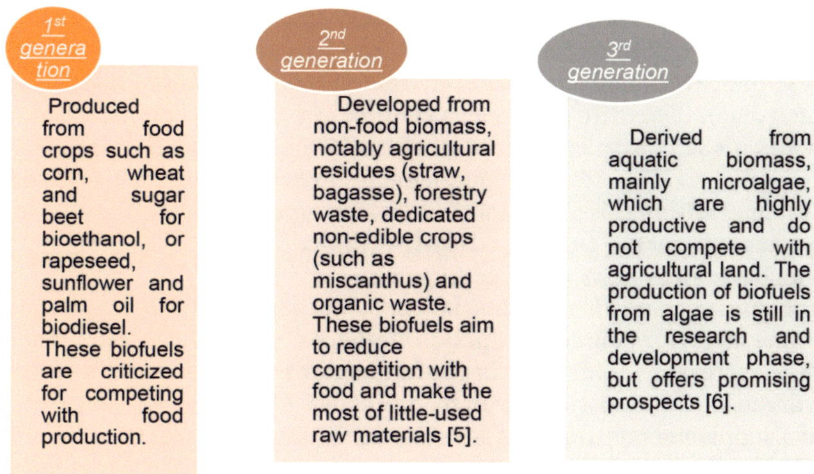


**Figure 1.**
*Evolution of biofuel generations: feedstocks and characteristics.*

in specific applications through enhanced catalytic conversion technologies that improve efficiency, integration into combined food-energy production systems that optimize land use, and strategic utilization of agricultural residues and waste materials rather than dedicated food crops [21–24]. Second-generation biofuels attempt to overcome food-versus-fuel and environmental challenges by utilizing lignocellulosic biomass such as agricultural residues, forestry waste, and energy crops [25, 26]. These feedstocks avoid direct competition with food production while providing opportunities for waste valorization. Lifecycle analyses suggest lower GHG emissions than fossil fuels due to carbon sequestration during growth [27]. Yet, commercialization remains difficult because lignocellulosic biomass is structurally resistant, necessitating costly pretreatments like steam explosion or acid hydrolysis [28]. Progress has been achieved through innovations in nanostructured enzymes and bifunctional catalysts [29]. While enzymatic hydrolysis and fermentation dominate current approaches, thermochemical methods such as pyrolysis and gasification show promise for diverse feedstocks [30]. Successful projects in Brazil, using sugarcane bagasse [31], and in Sweden, with forestry residues, demonstrate the importance of supportive policy frameworks [32]. Third-generation biofuels, derived from algae and photosynthetic microorganisms, represent a more sustainable alternative by avoiding food crop competition and requiring less arable land [33–36]. Algae can be cultivated in freshwater, seawater, or wastewater, achieving high productivity while absorbing carbon dioxide [37–39]. Their primary limitation lies in high production costs, driven by cultivation, harvesting, and processing complexities [40–43]. Advances in metabolic engineering, synthetic biology, and nanotechnology are being explored to enhance efficiency, strain performance, and energy return [44]. Despite challenges, algae-based fuels promise significant GHG reductions with minimal impact on food systems or land use, provided research and development efforts continue [45, 46].

## 2.2 The global energy context and biodiesel role in the energy transition

The global energy landscape is undergoing a profound transformation, driven by technological advancements, policy shifts, and growing environmental concerns.

According to the International Energy Agency (IEA) World Energy Outlook 2023, demand for traditional fossil fuels—coal, oil, and natural gas—is projected to peak before 2030 under the Stated Policies Scenario (STEPS). This anticipated milestone represents a major departure from over a century of fossil fuel dominance, during which these sources consistently supplied about 80% of global energy needs [47], highlighting the world's gradual but decisive shift towards diversified, low-carbon energy systems. Within this transition, biodiesel is emerging as a viable and strategically important alternative fuel, particularly for decarbonizing the transportation sector, which remains one of the largest contributors to global $CO_2$ emissions. Road transport alone accounts for a substantial portion of fossil fuel consumption, making it a prime target for emission reduction strategies. Research indicates that even a modest 10% increase in biodiesel's share of transport fuels could significantly slow the growth rate of $CO_2$ emissions from road vehicles. This positions biodiesel not merely as a supplementary fuel, but as an immediately deployable mitigation tool in the global effort to meet climate targets. The environmental performance of biodiesel, however, is highly dependent on the choice of feedstock. For instance, life cycle assessment (LCA) studies on rapeseed biodiesel generally report substantial reductions in greenhouse gas (GHG) emissions compared with conventional fossil diesel. However, the extent of these emission savings varies considerably depending on factors such as agricultural practices, fertilizer application, crop yield, and the energy intensity of biodiesel production and processing stages [48, 49]. This wide range of reported values indicates that the environmental performance of rapeseed biodiesel is highly dependent on how the feedstock is cultivated and converted. These results underscore the critical importance of feedstock selection and management in maximizing the environmental advantages of biodiesel. In addition, sustainability assessments should not be limited to emission reductions alone, but should also account for broader impacts, including land-use changes, soil and water resource management, and potential ecological consequences. Therefore, ensuring that biodiesel feedstocks are sourced from environmentally responsible and socially equitable supply chains is essential for biodiesel to effectively contribute to long-term clean energy and decarbonization strategies.

### 2.3 Limitations of food-based feedstocks and the case for non-food biomass

The production of first-generation biofuel, derived from food crops such as corn, sugarcane, soybean, and palm oil, has sparked significant debate due to its social, environmental, and economic implications. One of the most prominent concerns is the competition between fuel and food production. Diverting staple crops to fuel manufacturing can limit their availability for human consumption and livestock feed, driving up global food prices. Such price increases disproportionately affect low-income populations, particularly in developing nations where food expenditures account for a large share of household income [50]. Beyond economic impacts, the rising demand for biofuel feedstocks has contributed to large-scale land acquisitions, or "land grabbing," especially in parts of Africa, Asia, and Latin America. These transactions often involve foreign investors or corporations acquiring vast tracts of land for biofuel crop cultivation, displacing local farmers and indigenous communities. The result is not only the loss of livelihoods but also cultural disruption and increased social tensions [51]. Environmental drawbacks are also substantial. Expansion of agricultural land to meet biofuel feedstock demand has driven deforestation, habitat loss, and ecosystem degradation. Land-use change releases large amounts of stored carbon dioxide, potentially offsetting the climate benefits of biofuels. In countries like Indonesia and Brazil, clearing forests or draining peatlands for oil palm or sugarcane plantations has become a major source of GHG emissions [51].

Intensive monoculture farming associated with these crops also consumes significant water resources and relies heavily on fertilizers and pesticides, which can degrade soil quality, contaminate water supplies, and reduce long-term agricultural sustainability. Recognizing these limitations, many governments and international bodies have reassessed their biofuel strategies. Some have introduced caps or phase-out timelines for first-generation biofuels, while others are actively promoting alternatives with lower environmental and social risks. This has accelerated interest in non-food biomass as the foundation for second-generation biofuels. Such feedstocks include agricultural residues (e.g., straw, corn stover), forestry residues, organic municipal waste, and non-edible energy crops (**Figure 2**) [52]. Second-generation biofuels offer several advantages: they reduce competition for arable land, make use of abundant lignocellulosic materials that are otherwise underutilized, and enable the valorization of waste streams. Converting residues and organic waste into fuel not only reduces reliance on food crops but also supports circular economy principles by minimizing landfill use and lowering methane emissions from decomposition. Nonetheless, technical and economic barriers persist. Processing lignocellulosic biomass requires complex pretreatment to break down the rigid structure of plant cell walls, followed by enzymatic hydrolysis and fermentation. These steps can be costly and energy-intensive. However, advances in biotechnology, catalyst design, and process engineering are gradually improving conversion yields and reducing costs. With sustained research and development, the transition from food-based to non-food feedstocks could become both commercially viable and environmentally superior, reinforcing the role of biofuels as a sustainable component of the global energy transition.

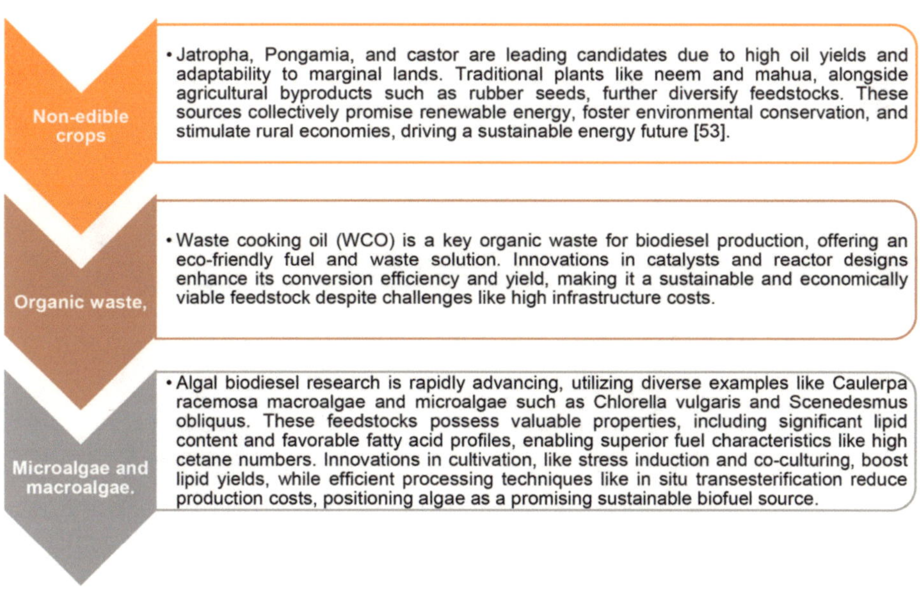


**Figure 2.**
*Typology of non-food biomass.*

## 3. Typology of non-food biomass

The systematic classification of non-food feedstocks for biodiesel production is a critical step in advancing sustainable bioenergy systems, particularly in the context of global energy transitions. Such classification enables a clearer understanding of feedstock availability, conversion potential, and environmental performance, while helping to minimize competition with food resources and reduce associated socio-economic and ecological impacts. These renewable resources are typically categorized into three groups: non-edible crops, organic waste, and microalgae/macroalgae (**Figure 2**). Each category presents distinct opportunities and challenges across the biofuel value chain. Rising global demand for renewable energy has stimulated interest in non-edible crops, which avoid competition with food resources. Unlike edible oils such as soybean or palm oil, these crops can yield substantial oil without compromising food security. Many non-edible crops are well-suited to marginal lands, requiring minimal water and fertilizers, enhancing both their economic viability and environmental sustainability [53]. Their oils contain fatty acids suitable for transesterification, producing biodiesel with properties comparable to conventional diesel. Non-edible biodiesel also reduces greenhouse gas emissions and enhances energy independence. Research continues to optimize oil extraction, engine performance, and cost efficiency, positioning non-edible crops as key players in the future of clean energy. These crops have gained attention as sustainable feedstocks, offering an eco-friendly alternative to conventional diesel without competing with food production. For example, *Jatropha curcas* (**Figure 3**) is particularly promising due to its high oil content and ability to thrive in arid conditions with minimal inputs [54]. Similarly, *Pongamia pinnata* provides comparable oil yields while improving soil fertility through natural nitrogen fixation (**Figure 3**) [55]. Castor plants offer high oil content but require careful handling due to toxic compounds. Traditional resources such as neem and mahua remain important where naturally abundant. Agricultural byproducts, like rubber seeds, also represent viable biodiesel sources. Collectively, these non-edible feedstocks support renewable energy production on marginal lands, promote

**Figure 3.**
*Jatropha as a non-edible feedstock for biodiesel production [54].*

environmental conservation, enhance energy independence, and create economic opportunities in rural areas. Their continued development mitigates reliance on fossil fuels and reduces the food-versus-fuel dilemma. *Jatropha curcas*, in particular, offers 27–40% oil content, tolerance to marginal lands, and no competition with food crops [54–58]. The plant's oil can be converted through transesterification into biodiesel meeting standard fuel specifications and international EN and ASTM standards [59]. Key advantages include its substantial oil yield, non-edible status that preserves food security, and capacity to grow in poor soil conditions with minimal water requirements [60–63]. Environmentally, Jatropha cultivation offers carbon sequestration benefits while reclaiming degraded lands [64–66]. However, challenges persist including high production costs from labor-intensive harvesting, seed toxicity requiring careful handling, and variable yields impacting economic viability [67–69]. Recent technological advances are addressing these limitations through improved transesterification processes using novel catalysts [70]. The crop demonstrates significant socioeconomic potential by creating rural employment opportunities and enhancing energy security, particularly in developing regions [71, 72]. Recent studies demonstrate that biodiesel derived from Pongamiapinnata, offers a viable alternative to fossil diesel, balancing engine performance with environmental benefits [73–79]. For example, Gaur and Goyal found that a 20% blend of mixed biodiesel (waste cooking oil and Pongamia oil) reduced CO and HC emissions by 30% while maintaining engine compatibility, though NOx emissions increased due to higher oxygen content [73]. Similarly, Mulla and Shah reported that Pongamia biodiesel produced via CaO nanocatalysts achieved 95% yield, with B20 blends showing a 5.7% increase in indicated power (**Figure 4**) [74]. Biocatalysts, such as banana waste-derived nanocatalysts, further optimized transesterification, yielding 96.3% biodiesel from Jatropha-Pongamia blends in just 10 minutes [75]. Dolomite-based catalysts also proved effective for Pongamia, achieving 92.3% yield while supporting afforestation and SDGs [76]. Coating engines with thermal barriers improved combustion efficiency for Pongamia-Azadirachta blends, reducing HC and CO emissions but increasing NOx by 9% [77]. Statistically optimized Pongamia biodiesel met ASTM standards, with a 78.3% yield under mild reaction

**Figure 4.**
*Pongamia tree flowering 1.5 years after planting in a trial plot in Buntoi, PulangPisau, Central Kalimantan [55].*

conditions (55°C, 60 minutes) [78]. Lemongrass oil (LGO) has emerged as a promising non-edible biodiesel feedstock, demonstrating competitive engine performance and emission characteristics compared to conventional diesel (**Figure 5**) [79]. Experimental studies on 3.5–5.2 kW diesel engines reveal that LGO blends (10–30% v/v) achieve brake thermal efficiency (BTE) comparable to diesel (LG20 ≈ diesel) but exhibit higher specific fuel consumption (SFC) due to lower calorific value [80]. Emission analyses show significant reductions in nitrogen and carbon oxides [80–82], attributed to LGO's oxygenated combustion and lower peak flame temperatures. Advanced formulations, such as LGO biodiesel with dibutyl ether additive, further optimize performance at compression ratio 17.5, reducing CO, $CO_2$, and NOx while maintaining mechanical efficiency [83]. Recent studies have explored novel approaches that incorporate bio-based multi-wall carbon nanotubes into lemongrass oil (LGO) emulsions, revealing multiple performance enhancements. Specifically, this addition has been shown to significantly reduce the ignition delay by 17.6% and shorten the overall combustion duration by 18.5% [84]. These improvements are primarily attributed to the superior thermal conductivity of additive, which promotes more efficient heat transfer within the fuel droplets, leading to faster vaporization, improved atomization, and more complete combustion. Consequently, integrating BMWCNTs into LGO emulsions not only enhances combustion efficiency but may also contribute to reduced pollutant formation and better engine performance.

Another promising non-edible biodiesel source is neem (*Azadirachta indica*) oil, which contains 30–40% oil in its seeds and can be converted to biodiesel through transesterification [85, 86]. Optimizing reaction parameters, such as temperature, methanol-to-oil ratio, and catalyst concentration, is crucial to achieving high biodiesel yield [86]. The resulting neem biodiesel exhibits fuel properties, including density, kinematic viscosity, and calorific value, that comply with ASTM standards, making it suitable for diesel engines [87, 88]. Studies on biodiesel blends enhanced with alumina nanoparticles demonstrate improved combustion efficiency and reduced emissions of CO, HC, and particulate matter [89]. Machine learning models have been applied to predict engine performance and emission behavior with high accuracy, reinforcing the feasibility of neem biodiesel as a sustainable fuel [89]. Additionally, neem butyl ester produced via acid-catalyzed transesterification has shown comparable performance and emission characteristics to conventional diesel in compression ignition engines [90]. Furthermore, the use of organic waste-derived feedstocks, such as waste cooking oil, provides an eco-friendly and low-cost

**Figure 5.**
*Lemongrass plant (A), lemongrass oil (B), and lemongrass stalks (C) [79].*

alternative for biodiesel production, aligning with circular economy principles [89, 91]. Contemporary research has significantly advanced waste cooking oil (WCO) biodiesel production through innovative reactor designs and catalytic processes. Jarungwongsathien et al. demonstrated the efficacy of a rotating tube reactor (RTR) system, achieving 89.72% biodiesel yield in just 11.23 minutes while reducing free fatty acid (FFA) content from 6 to 1.89% - with economic analysis showing a remarkably short 3.1-year payback period due to reduced operating costs [92]. Catalyst development has been particularly transformative, as shown by Alahmar et al. who engineered a bifunctional Ce/Mn/$\gamma$-$Al_2O_3$ catalyst that attained 97% conversion of high-FFA WCO through simultaneous esterification and transesterification, leveraging its high basicity (1.543 mmol/g) and surface area (143 $m^2$/g) [93]. Further innovations in nanocatalysts include Rezaeifar et al.'s $\alpha Fe_2O_3$/CuO composite RSM optimization showed the methanol/WCO ratio was the most impactful variable yielding a 94.27% biodiesel yield under optimal conditions. The catalyst exhibited strong reusability, demonstrating its potential as an effective heterogeneous catalyst [94]. Green chemistry approaches have also emerged, with Sahu et al. synthesized a heterogeneous CaO nanocatalyst using watermelon peel extract for the transesterification of soybean oil into biodiesel. Optimal conditions (20:1 methanol:oil ratio, 8% catalyst, 2.5 h at 65°C) resulted in a high biodiesel yield [95]. Selvam et al. evaluated pentanol-blended WCO-biodiesel. The B30+ blend demonstrated superior performance compared to pure diesel, improving combustion (higher heat release and pressure). Crucially, this blend achieved substantial reductions in CO (31.75%), HC (20.28%), and NO (14.82%) compared to other biodiesel blends, proving it to be a sustainable and highly efficient automotive fuel [96]. The integration of artificial intelligence has further optimized performance parameters, as Biberci et al. investigation utilized artificial neural network (ANN) and Response Surface Methodology models to predict and optimize the performance and exhaust emissions of a diesel engine fueled by a blend of diesel and waste oil biodiesel. The RSM model achieved high accuracy successfully maximizing brake thermal efficiency (BTE) while minimizing consumption and emissions. Optimal engine operation was achieved at an injection pressure of 180 bar and 100 percent biodiesel ratio, demonstrating that the combined use of RSM, ANN, and advanced algorithms offers a highly accurate optimization framework [97]. Comparative feedstock analyses by Liew et al. using non-edible oils such as soapnut seed kernel oil (SKO) and used cooking oil were studied [98]. SKO was extracted via Soxhlet at varying durations and characterized for pH, density, and acid number. Transesterification produced biodiesel, with SKO outperforming UCO, showing suitable pH (6.5), density (780 kg/$m^3$), and acid number (2.43 mg KOH/g) [98].

However, infrastructure costs, particularly for bio-refineries, remain a major barrier to large-scale biodiesel production. Innovative feedstock strategies, such as integrating waste cooking oil with microalgae, have shown promise in enhancing process efficiency and reducing overall production costs [99, 100]. Dejene demonstrated that hybrid biodiesel feedstocks combining microalgae and WCO can achieve high conversion efficiencies while maintaining desirable fuel properties, suggesting a feasible approach for sustainable biodiesel supply chains [100]. These collective advancements position WCO biodiesel as a technically mature alternative, though challenges persist in NOx mitigation, cold flow properties, and scaling heterogeneous catalyst production [92–99]. Recent studies demonstrate significant progress in algal biodiesel production through innovative cultivation and processing techniques [100]. Aboueldahab et al. [101] analyzes eight Red Sea seaweed species for biodiesel potential. Seasonal biochemical analysis revealed significant variations in lipids, carbohydrates, and proteins. *Caulerpa racemosa* was the most promising, with high, consistent lipid yields (up to 12.76% CDW) and a favorable fatty acid profile rich in saturated fats like palmitic acid. Predictive models indicated excellent fuel properties, including high cetane numbers and suitable cold-flow characteristics, confirming the seaweed's strong potential as a sustainable biodiesel source [101]. For microalgae, the comparison of *Chlorella vulgaris* biodiesel proves that the in situ transesterification reduced capital costs by 31% ($3.48 M vs. $5.05 M) and achieved a shorter payback period (1.72 years) compared to solvent extraction [102]. Engine performance studies reveal promising results: Allasi et al. showed *Scenedesmus obliquus* biodiesel with 20% n-butanol additive improved brake thermal efficiency while reducing CO and hydrocarbon emissions though NOx increased due to higher combustion temperatures [103]. Supply chain optimization is critical for scalability, as demonstrated by Panwar et al. , whose genetic algorithm model reduced costs by prioritizing facility placement and utilizing mixed feedstocks (macroalgae/WCO), though production facilities [104]. Life cycle assessments by Liu et al. revealed wastewater-cultivated microalgae reduced environmental impacts by 52% compared to freshwater systems, with carbon sequestration potential during cultivation offsetting 69.8% of supply chain emissions [105].

Sun et al. [106] tested *Dunaliella salina* microalgae biodiesel blends (DSB10-DSB30) in a diesel engine. Brake thermal efficiency was highest for pure diesel (DSB0), but biodiesel blends showed better performance between 2000 and 2500 rpm due to higher oxygen content, enhancing combustion. However, specific fuel consumption increased with biodiesel concentration. Emissions analysis revealed lower CO but higher $CO_2$ and NOx for biodiesel blends, particularly at higher speeds due to increased combustion temperatures [106]. Advanced cultivation techniques further improve yields. Poh et al. [107] investigated post-harvest stress on *Chlorella vulgaris* to increase lipid content for biodiesel. Applying one day of nutrient starvation combined with salinity (6.0 g/L) in darkness achieved the highest lipid yield of 38.8–40.28% by dry weight. The extracted lipids contained suitable fatty acids (linoleic, linolenic, and palmitic acid), similar to conventional vegetable oils, confirming their viability as a biodiesel feedstock [107]. Co-culture systems offer additional gains. By combining *Chlorella pyrenoidosa* and Cutaneotrichosporonoleaginosum, Wu et al. [108] achieved a lipid concentration of 18.15 g/L and a lipid productivity of 4.03 g/L/day. The system enabled pH self-regulation and nutrient synergism, scaling effectively even with cassava starch hydrolysate and demonstrating pH self-regulation and nutrient synergism [108].

## 4. Technologies for converting non-food biomass into biodiesel

The shift from fossil fuels to renewable energy has accelerated the development of technologies for converting non-food biomass into biodiesel. While the core chemical process of transesterification remains constant, the methods of its execution have evolved significantly to address the limitations of traditional approaches and improve overall sustainability and economic viability. This evolution has led to the development of advanced processes utilizing heterogeneous catalysts, microwave, and ultrasonic technologies, each offering distinct advantages over the classic use of homogeneous catalysts. Historically, homogeneous

*DOI: http://dx.doi.org/10.5772/intechopen.1012287*

catalysts, such as sodium hydroxide and potassium hydroxide, have been the standard for biodiesel production. These catalysts dissolve completely in the reaction mixture, ensuring excellent contact with the reactants and leading to fast reaction rates and high conversion yields. However, this method has significant drawbacks. The main challenge lies in the complex and energy-intensive downstream processing required to separate the catalyst and purify the biodiesel. This typically involves multiple washing steps, which not only consume a lot of energy but also generate a large volume of contaminated wastewater, creating environmental and disposal challenges. Furthermore, since these catalysts cannot be recovered and reused, the process becomes less economically and environmentally sustainable in the long run. The limitations of homogeneous catalysts led to a strong push for heterogeneous catalysts. These are solid catalysts that do not dissolve in the reaction mixture, which simplifies the separation process immensely. They can be easily removed by filtration or centrifugation and, most importantly, can be reused, reducing operational costs and waste generation. This reusability also facilitates continuous-flow production systems, making the technology more scalable for industrial applications. Recent research has yielded remarkable results with heterogeneous catalysts [109–124]. For example, Santoso et al. [111] optimized biodiesel synthesis from waste cooking oil using a $Na_2O/\gamma$-$Al_2O_3$ catalyst and ultrasonic assistance. The best yield (83.51%) was achieved with a 3% catalyst loading, a 12:1 methanol-to-oil molar ratio, at 65°C for 15 minutes. The main biodiesel components identified were methyl elaidate, methyl palmitate, and methyl linoleate. The catalyst's crystal size was 30.59 nm, and the process successfully reduced free fatty acids for efficient transesterification [111]. The development of green synthesis methods has also emerged, with researchers successfully creating calcium oxide (CaO) nanoparticles from waste orange peel extract to achieve a 93.4% biodiesel yield from WCO [113]. Other notable examples include a nano-magnetic $CaO/Fe_2O_3$/feldspar catalyst that provides high yields from various wild plant seed oils, with the added benefit of easy magnetic separation and reusability [118]. A significant breakthrough is the creation of catalysts, like hierarchical porous $\gamma$-$Al_2O_3$ modified with MoV metal oxides, that can perform simultaneous esterification and transesterification in a single step, which simplifies the process and enhances cost-effectiveness [123]. Microwave-assisted transesterification is essential for process efficiency. Unlike conventional heating, which heats from the outside in, microwaves directly and rapidly heat the polar molecules in the reaction mixture. This targeted and instantaneous heating dramatically accelerates reaction kinetics, reducing reaction times from hours to mere minutes [124]. This technology offers improved energy efficiency by minimizing heat loss. It is particularly effective for in situ transesterification of microalgae, as it combines oil extraction and conversion into a single step, which is ideal for large-scale, sustainable biofuel production. The use of ultrasonic waves is another powerful method of process intensification. Ultrasound operates on the principle of acoustic cavitation, where sound waves create and collapse microscopic bubbles in the liquid. This violent collapse generates localized high temperatures and pressures, effectively breaking down the immiscible oil and alcohol phases and significantly enhancing mass transfer between the reactants. This physical action leads to a considerable increase in reaction rate and efficiency. Studies have shown that ultrasonic-assisted transesterification can achieve high yields while also reducing energy consumption by up to 85% compared to conventional methods [125–128].

## 5. Technological and environmental challenges

Feedstock variability poses a dual challenge and opportunity for the widespread adoption of biodiesel derived from non-food biomass. Unlike petroleum, a relatively uniform raw material, non-food biomass is a diverse and inconsistent collection of materials, including dedicated oilseed crops, agricultural residues, and organic wastes. This variability is a defining characteristic of the bioenergy sector, impacting every stage of production from collection and pretreatment to the final quality of the biodiesel. The primary challenge stems from compositional differences, such as variations in oil content, fatty acid profiles, and impurity levels (e.g., free fatty acids and moisture) across different feedstocks. These variations can be influenced by species, growing conditions, and storage duration. For example, a feedstock with a high free fatty acid content may require a two-step transesterification process to prevent soap formation, whereas a different feedstock with lower FFA levels might be processed using a single step [129]. This lack of uniformity in feedstock composition directly hinders the standardization of conversion processes. A process optimized for a specific type of biomass may be inefficient or fail when applied to another, leading to increased capital and operational costs. Facilities must be flexible enough to handle various feedstocks, requiring more related to feedstock variability in biodiesel production.

## 6. Summary of key points

This chapite [part 1] highlights for the critical biodiesel production role in transitioning towards a sustainable energy future. The findings presented across various sections of this paper demonstrate that non-food feedstocks—including dedicated crops, agricultural wastes, and algae—offer a compelling solution to the "food-versus-fuel" dilemma, while simultaneously providing a pathway for waste valorization. Technological advancements in transesterification have been instrumental in this progress. The shift from classic, environmentally problematic homogeneous catalysts to more efficient and sustainable heterogeneous catalysts has significantly improved biodiesel yields and simplified purification processes. Modern techniques like microwave and ultrasonic assisted transesterification further intensify these processes, drastically reducing reaction times and energy consumption. These innovations, combined with the strategic application of artificial intelligence and advanced modeling, are making non-food biodiesel a technically viable alternative to fossil fuels. However, several key challenges remain that must be addressed for widespread adoption. The variability in feedstock composition requires flexible and robust processing technologies to ensure consistent biodiesel quality. The economics of production are heavily dependent on optimizing the entire value chain, from feedstock sourcing and collection to the valorization of co-products in an integrated biorefinery model. Furthermore, the environmental impacts of biodiesel must be continuously evaluated using life cycle assessments to ensure that benefits such as reduced greenhouse gas emissions are not offset by negative effects like increased NOx emissions or unsustainable land use. Finally, the logistical and infrastructural challenges of building a new supply chain for a diverse set of biomass feedstocks remain a significant hurdle. In summary, the transition to non-food biodiesel is a complex but necessary journey. With continued research and development focused on innovative catalysts, efficient processing technologies, and integrated, circular-economy-based supply chains, non-food biodiesel is well-positioned to become a cornerstone of the global energy transition, providing a renewable, cleaner, and more sustainable fuel for the future.

**Acknowledgements**

The authors gratefully acknowledge the support received from the PYRAGRAF project, financed under the European Union's Horizon Europe framework (Grant Agreement ID 101114608), which provided essential insights into decentralized pyrolytic conversion of agricultural and forestry residues. We also extend our gratitude to the SALTOpower project, financed under the European Union's Horizon Europe framework (GrantAgreement ID 101079303), coordinated by the University of Évora, for their contributions on molten salt-based energy storage and its integration with biomass pyrolysis-drivenhydrogen/syngas systems. The authors acknowledge the use of artificial intelligence tools to enhance the clarity, grammar, and academic style of the text. All scientific content, analyses, interpretations, and conclusions presented in this chapter were independently developed, verified, and validated by the authors, ensuring the accuracy, integrity, and originality of the work.

## Conflict of interest

The authors declared no conflicts of interest.

## Author details

Asma Nouira[1*], Imen Amri[1], Imene Bekri-Abbes[2] and Diogo Canavarro[1]

1 Renewable Energies Chair, Polo da Mitra da Universidade de Évora, Nossa Senhora da Tourega, Portugal

2 National Center of Research of Sciences of Materials, Technopole of Borj Cedria, Tunis, Tunisia

*Address all correspondence to: asma.nouira@uevora.pt and asma.nouira@fst.utm.tn

## References

[1] Capgemini. Biofuels: Potential, applications, and challenges. 2024. Available from: https://www.capgemini.com/insights/expert-perspectives/biofuels-potential-applications-and-challenges/ [Accessed: May 13, 2025]

[2] Parameswaran B, Gnansounou E, Raveendran S, Pandey A. Enzymes for second generation biofuels: Recent developments and future perspectives. Bioresource Technology Reports. 2018;**5**:52-61. DOI: 10.1016/j.biteb.2018.06.005

[3] Biofuels Offer Hope for Hard-to-Decarbonise Sectors. The Australian. Available from: https://www.theaustralian.com.au/business/mining-energy/biofuels-offer-hope-for-hardtodecarbonise-sectors/news-story/81c37a6602760947a29e6c7d72c91bb1; [Accessed: May 13, 2025]

[4] Cultures Sciences Chimie. Les biocarburants : de la 1ère à la 3ème génération. Disponible sur: https://culturesciences.chimie.ens.fr/thematiques/chimie-organique/catalyse/les-biocarburants-de-la-1ere-a-la-3eme-generation [Consulted: Mai 13, 2025]

[5] Ministère de l'Agriculture. Tout savoir sur les biocarburants. Disponible sur: https://agriculture.gouv.fr/tout-savoir-sur-les-biocarburants [Consulted: Mai 13, 2025]

[6] Ecosources. Les biocarburants de troisième génération. Disponible sur: https://www.ecosources.org/biocarburant-de-troisieme-generation [Consulted: Mai 13, 2025]

[7] Deshmukh A. Technical and economical feasibility of biofuels for engine applications. In: SAE Technical Paper 2009-24-0087. Warrendale (PA): Society of Automotive Engineers; 2009. DOI: 10.4271/2009-24-0087

[8] Vilas Bôas RN, Mendes MF. A review of biodiesel production from non-edible raw materials using the transesterification process with a focus on influence of feedstock composition and free fatty acids. Journal of the Chilean Chemical Society. 2022;**67**(1):5433-5444. DOI: 10.4067/S0717-97072022000105433

[9] Makarevičienė V, Gaidė I, Sendžikienė E. Heterogeneous catalysts from food waste for biodiesel synthesis—A comprehensive review. Catalysts. 2025;**15**(10):957. DOI: 10.3390/catal15100957

[10] Wengenmayr R. Renewable Energy: Sustainable Concepts for the Energy Change. 2nd ed. Weinheim: Wiley-VCH; 2013

[11] Adessi A, De Philippis R. Biodiesel from microalgae. In: Hallenbeck PC, editor. Microbial Technologies in Advanced Biofuels Production. Springer; 2012. DOI: 10.1007/978-1-4614-1208-3_4

[12] Mahapatra S, Kumar D, Singh B, Sachan PK. Biofuels and their sources of production: A review on cleaner sustainable alternative against conventional fuel, in the framework of the food and energy nexus. Energy Nexus. 2021;**4**:100036. DOI: 10.1016/j.nexus.2021.100036

[13] Nowack KF. Biofuels today and tomorrow. Hydrocarbon Process. 2007;**86**(2):87-95

[14] Ackrill R, Kay A. Biofuels policy challenges. In: Ackrill R, Kay A,

editors. The Growth of Biofuels in the 21st Century, Energy, climate and the environment series. London: Palgrave Macmillan; 2014. pp. 180-217. DOI: 10.1057/9781137307897_9

[15] Haque SM, Bhat AH, Khan I. Biomass: An ageless raw material for biofuels. In: Kaur GH, Khan I, Bhat AH, editors. Agricultural Biomass Based Potential Materials. Springer; 2015. pp. 442-466. DOI: 10.1007/978-3-319-13847-3_20

[16] Zhang Y, Dube MA, McLean DD, Kates M. Biodiesel production from waste cooking oil: 2. Economic assessment and sensitivity analysis. Bioresource Technology. 2003;**90**(3):229-240. DOI: 10.1016/s0960-8524(03)00150-0

[17] Searchinger T, Heimlich R, Houghton RA, Dong F, Elobeid A, Fabiosa J, et al. Use of U.S. croplands for biofuels increases greenhouse gases through emissions from land-use change. Science. 2008;**319**(5867):1238-1240. DOI: 10.1126/science.1151861

[18] European Commission. Renewable Energy – Recast to 2030 (RED II). EU Science Hub; 2018. Available from: https://joint-research-centre.ec.europa.eu/welcome-jec-website/reference-regulatory-framework/renewable-energy-recast-2030-red-ii_en

[19] US Environmental Protection Agency. Overview of the Renewable Fuel Standard Program. EPA; 2025. Available from: https://www.epa.gov/renewable-fuel-standard-program/program-overview-renewable-fuel-standard-program

[20] US Environmental Protection Agency. Final Renewable Fuels Standards Rule for 2023, 2024, and 2025. EPA; 2023. Available from: https://www.epa.gov/renewable-fuel-standard/final-renewable-fuels-standards-rule-2023-2024-and-2025

[21] Baskar G, Aiswarya R. Trends in catalytic production of biodiesel from various feedstocks. Renewable and Sustainable Energy Reviews. 2016;**57**:496-504. DOI: 10.1016/j.rser.2015.12.101

[22] Rezaei R, Mohadesi M, Moradi GR. Optimization of biodiesel production using waste mussel shell catalyst. Fuel. 2013;**109**:534-541. DOI: 10.1016/j.fuel.2013.03.004

[23] Yaghi M, Chidiac S, Awad S, El Rayess Y, Zgheib N. An overview of biodiesel production via heterogeneous catalysts: Synthesis, current advances, and challenges. Clean Technologies. 2025;**7**(3):62. DOI: 10.3390/cleantechnol7030062

[24] Meriatna HH, Riza M, Faisal M, Ahmadi S. Biodiesel production using waste banana peel as renewable base catalyst. Materials Today Proceedings. 2023;**87**(Pt. 2):214-217. DOI: 10.1016/j.matpr.2023.02.400

[25] Lee RA, Lavoie JM. From first- to third-generation biofuels: Challenges of producing a commodity from a biomass of increasing complexity. Animal Frontiers. 2013;**3**(2):6-11. DOI: 10.2527/af.2013-0010

[26] Kowalski Z, Kulczycka J, Verhé R, Desender L, De Clercq G, Makara A, et al. Second-generation biofuel production from the organic fraction of municipal solid waste. Frontiers in Energy Research. 2022;**25**:919415. DOI: 10.3389/fenrg.2022.919415

[27] Pramanik A, Sinha A, Chaubey KK, Hariharan S, Dayal D, Bachheti RK, et al. Second-generation bio-fuels:

Strategies for employing degraded land for climate change mitigation meeting united nation-sustainable development goals. Sustainability. 2023;**15**(9):7578. DOI: 10.3390/su15097578

[28] Figueiredo M, Mendes F. Problems found in the adaptation of a fluid cracking catalytic pilot plant for studying second generation biofuel production by pyrolysis. Chemical Engineering Transactions. 2012;**29**:691-696. DOI: 10.3303/CET1229116

[29] Jintao H, MuhdNoor ND, Kamarudin NHA, Ali MSM. Review of enzyme catalysis in sustainable biofuel production: A comparative study of waste oils and alternative feedstocks. International Journal of Green Energy. 2025;**22**(1):1-15

[30] Tagade A, Kirti N, Sawarkar AN. Pyrolysis of agricultural crop residues: An overview of researches by Indian scientific community. Bioresource Technology Reports. 2021;**15**:100761. DOI: 10.1016/j.biteb.2021.100761

[31] Rossi LM, Gallo JMR, Mattoso LHC, Buckeridge MS, Licence P, Allen DT. Ethanol from sugarcane and the Brazilian biomass-based energy and chemicals sector. ACS Sustainable Chemistry & Engineering. 2021;**9**(12):4293-4295. DOI: 10.1021/acssuschemeng.1c01678

[32] Yacout DMM, Tysklind M, Upadhyayula VKK. Socio-economic implications of forest-based biofuels for marine transportation in the Arctic: Sweden as a case study. Frontiers in Climate. 2024;**6**:1414813. DOI: 10.3389/fclim.2024.1414813

[33] Alam F, Mobin S, Chowdhury H. Third generation biofuel from algae. Procedia Engineering. 2015;**105**:763-768. DOI: 10.1016/j.proeng.2015.05.068

[34] Behera S, Singh R, Arora R, Sharma NK, Shukla M, Kumar S. Scope of algae as third generation biofuels. Frontiers in Bioengineering and Biotechnology. 2015;**2**:90. DOI: 10.3389/fbioe.2014.00090

[35] Dragone G, Fernandes BD, Vicente AA, Teixeira JA. Third generation biofuels from microalgae. In: Méndez-Vilas A, editor. Current Research, Technology and Education Topics in Applied Microbiology and Microbial Biotechnology. 2010. pp. 1355-1366

[36] Lackner M, Köpke M. Third-generation biofuels: Bacteria and algae for better yield and sustainability. In: Lackner M, Sajjadi B, Chen W-Y, editors. Handbook of Climate Change Mitigation and Adaptation. 3rd ed. Springer; 2025. pp. 2411-2466. DOI: 10.1007/978-3-031-84483-6_90

[37] Li G, Yao J. A review of algae-based carbon capture, utilization, and storage (algae-based CCUS). Gases. 2024;**4**(4):376-398. DOI: 10.3390/gases4040024

[38] Tang Y. Algae used in achieving carbon neutrality by treating wastewater. International Journal of Energy. 2025;**8**(1):61-66. DOI: 10.54097/v8tjht95

[39] Hoque MM, Iannelli V, Padula F, Radice RP, Saha BK, Martelli G, et al. Microalgae: Green engines for achieving carbon sequestration, circular economy, and environmental sustainability—A review based on last ten years of research. Bioengineering. 2025;**12**(9):909. DOI: 10.3390/bioengineering12090909

[40] Veal MW, Caffrey KR, Chinn MS, Grunden AM. Algae for Biofuels – Economic and Environmental Costs. SRAC Publication No. 4310. Southern Regional Aquaculture Center, Mississippi

State University; 2013. Available from: https://srac.msstate.edu/pdfs/Fact%20Sheets/4310%20Algae%20for%20Biofuels-%20Economic%20and%20Environmental%20Costs.pdf

[41] Hannon M, Gimpel J, Tran M, Rasala B, Mayfield S. Biofuels from algae: Challenges and potential. Biofuels. 2010;**1**(5):763-784. DOI: 10.4155/bfs.10.44

[42] Adeniyi OM, Azimov U, Burluka A. Algae biofuel: Current status and future applications. Renewable and Sustainable Energy Reviews. 2018;**90**:316-335. DOI: 10.1016/j.rser.2018.03.067

[43] Papagianni S, Capellán-Pérez I, Adam A, Pastor A. Review and meta-analysis of energy return on investment and environmental indicators of biofuels. Renewable and Sustainable Energy Reviews. 2024;**203**:114737. DOI: 10.1016/j.rser.2024.114737

[44] Rial RC. Biofuels versus climate change: Exploring potentials and challenges in the energy transition. Renewable and Sustainable Energy Reviews. 2024;**196**:114369. DOI: 10.1016/j.rser.2024.114369

[45] Kant G, Hasan A, Yadav P, Pandey A, Srivastava S. The generational shift in biofuels: A path toward sustainable energy solutions. Biomass and Bioenergy. 2025;**196**:107757. DOI: 10.1016/j.biombioe.2025.107757

[46] Chowdhury R, Caetano N, Franchetti MJ, Hariprasad K. Life cycle based GHG emissions from algae based bioenergy with a special emphasis on climate change indicators and their uses in dynamic LCA: A review. Sustainability. 2023;**15**(3):1767. DOI: 10.3390/su15031767

[47] International Energy Agency. World Energy Outlook 2023. Paris: IEA; 2023. DOI: 10.1787/827374a6-en

[48] Yang X, Liu Y, Bezama A, Thrän D. Two birds with one stone: A combined environmental and economic performance assessment of rapeseed-based biodiesel production. GCB Bioenergy. 2022;**14**(2):215-241. DOI: 10.1111/gcbb.12913

[49] Gupta R, McRoberts R, Yu Z, Smith C, Sloan W, You S. Life cycle assessment of biodiesel production from rapeseed oil: Influence of process parameters and scale. Bioresource Technology. 2022;**360**:127532. DOI: 10.1016/j.biortech.2022.127532

[50] Fletcher RJ Jr, Robertson BA, Evans J, Doran PJ, Alavalapati JRR, Schemske DW. The impacts of biofuel crops on local biodiversity: A global synthesis. Biodiversity and Conservation. 2021;**30**(10):2863-2883

[51] Abreu M, Silva L, Ribeiro B, Ferreira A, Alves L, Paixão SM, et al. Low indirect land use change (ILUC) energy crops to bioenergy and biofuels—A review. Energies. 2022;**15**(12):4348. DOI: 10.3390/en15124348

[52] Antizar-Ladislao B, Turrion-Gomez JL. Second-generation biofuels and local bioenergy systems. Biofuels Bioproducts Biorefining. 2008;**2**(5):455-469. DOI: 10.1002/bbb.97

[53] Winberg J, Ekroos J, Eklundh L, Smith HG. Constraints on the availability of marginal land for bioenergy production in southern Sweden. Biomass and Bioenergy. 2024;**190**:107421. DOI: 10.1016/j.biombioe.2024.107421

[54] Neupane D, Bhattarai D, Ahmed Z, Das B, Pandey S, Solomon JKQ, et al. Growing Jatropha (*Jatropha curcas* L.) as a potential second-generation biodiesel feedstock. Inventions. 2021;**6**(4):60. DOI: 10.3390/inventions6040060

[55] Leksono B, Rahman SA, Larjavaara M, Purbaya DA, Arpiwi NL, Samsudin YB, et al. Pongamia: A possible option for degraded land restoration and bioenergy production in Indonesia. Forests. 2021;**12**(11):1468. DOI: 10.3390/f12111468

[56] Ruatpuia JVL, Halder G, Vanlalchhandama M, Lalsangpuii F, Boddula R, Al-Qahtani N, et al. *Jatropha curcas* oil as a potential feedstock for biodiesel production: A critical review. Fuel. 2024;**370**:131829. DOI: 10.1016/j.fuel.2024.131829

[57] Yusuf NNAN, Kamarudin SK, Yaakob Z. Overview on the production of biodiesel from *Jatropha curcas* L. by using heterogeneous catalysts. Biofuels, Bioproducts and Biorefining. 2012;**6**(3):319-334. DOI: 10.1002/bbb.345

[58] Che Hamzah NH, Khairuddin N, Siddique BM, Hassan MA. Potential of *Jatropha curcas* L. as biodiesel feedstock in Malaysia: A concise review. PRO. 2020;**8**(7):786. DOI: 10.3390/pr8070786

[59] Sakthivel R, Ramesh K, Purnachandran R, Mohamed SP. A review on the properties, performance and emission aspects of the third generation biodiesels. Renewable and Sustainable Energy Reviews. 2018;**82**:2970-2992. DOI: 10.1016/j.rser.2017.10.037

[60] Rodrigues J, Miranda I, Furquim L, Gominho J, Vasconcelos M, Barradas G, et al. Storage stability of *Jatropha curcas* L. oil naturally rich in gamma-tocopherol. Industrial Crops and Products. 2015;**64**:188-193. DOI: 10.1016/j.indcrop.2014.10.048

[61] Abou Kheira AA, Atta NMM. Response of *Jatropha curcas* L. to water deficits: Yield, water use efficiency and oilseed characteristics. Biomass and Bioenergy. 2009;**33**(10):1343-1350. DOI: 10.1016/j.biombioe.2008.05.015

[62] Pérez-Vázquez A, Hernández-Salinas G, Ávila-Reséndiz C, Valdés-Rodríguez OA, Gallardo-López F, García-Pérez E, et al. Effect of the soil water content on Jatropha seedlings in a tropical climate. International Agrophysics. 2013;**27**(4):351-357. DOI: 10.2478/intag-2013-0004

[63] Al-Busaidi A, Ahmed M, Chikara J. The impact of heat and water stress conditions on the growth of the biofuel plant *Jatropha curcas*. International Journal of Environmental Studies. 2012;**13**(1):273-288. DOI: 10.1080/00207233.2012.663204

[64] Wani SP, Chander G, Sahrawat KL, Srinivasa Rao C, Raghvendra G, Susanna P, et al. Carbon sequestration and land rehabilitation through *Jatropha curcas* (L.) plantation in degraded lands. Agriculture, Ecosystems and Environment. 2012;**161**:112-120. DOI: 10.1016/j.agee.2012.07.028

[65] Baumert S, Khamzina A, Vlek PLG. Soil organic carbon sequestration in *Jatropha curcas* systems in Burkina Faso. Land Degradation and Development. 2016;**27**(6):1814-1823. DOI: 10.1002/ldr.2310

[66] Zhang Z, Li J, Wang Z, Liu H, Wei K. Estimating soil carbon sequestration of Jatropha for sustainable aviation fuel pathway. Water, Air, and Soil Pollution. 2024;**235**(1):47. DOI: 10.1007/s11270-023-06826-5

[67] Segerstedt A, Bobert J. Revising the potential of large-scale Jatropha oil production in Tanzania: An economic land evaluation assessment. Energy Policy. 2013;**57**:491-505. DOI: 10.1016/j.enpol.2013.02.023

[68] Mehra KS, Singh S, Singh AK, Kharkwal H, Avikal S, et al. Performance, energy, emission and cost analysis of Jatropha (*Jatropha Curcas*) oil as a biofuel for compression ignition engine. Materials Today Proceedings. 2021;**43**(Part 1):348-354. DOI: 10.1016/j.matpr.2020.11.675

[69] Liu Y, Zhu Z, Zhang R, Zhao X. Life cycle assessment and life cycle cost analysis of Jatropha biodiesel production in China. Biomass Conversion and Biorefinery. 2024;**14**(22):28635-28660

[70] Shafi ME, Alsabi HA, Almasoudi SH, Mufti FAM, Alowaidi SA, Alaswad AA. Catalytic conversion of *Jatropha curcas* oil to biodiesel using mussel shell-derived catalyst: Characterization, stability, and comparative study. Inorganics. 2024;**12**(4):109. DOI: 10.3390/inorganics12040109

[71] Yang L, Zhou J, Pan C, Tu Q, Wang G, Duan J, et al. Environmental, energy, and economic assessment of *Jatropha curcas* L. biodiesel production in China. Journal of Cleaner Production. 2024;**470**:143286. DOI: 10.1016/j.jclepro.2024.143286

[72] Nahar K. *Jatropha curcas* L: A Potential 2G Energy Crop to Produce Biofuel in Bangladesh: Agronomy, Biotechnology, Biodiesel and Byproducts. Cham: Springer Nature Switzerland; 2025

[73] Gaur RK, Goyal R. Performance, emission, combustion and economics analysis of CI engine fueled with mixed oil biodiesel blends from waste cooking oil and pongamia oil. SN Applied Sciences. 2025;**7**(6):365. DOI: 10.1007/s42452-025-07150-7

[74] Mulla AA, Shah AP. Production and engine performance analysis of biodiesel from *Pongamia pinnata* using PTSA and CaO nanocatalysts. Biomass and Bioenergy. 2025;**195**:107714. DOI: 10.1016/j.biombioe.2024.107714

[75] Basumatary SF, Das B, Brahma S, Basumatary S. Musa ABB (Kachkal) banana waste derived heterogeneous nanocatalyst for transesterification of binary oil mixture of *Jatropha curcas* and *Pongamia pinnata* to biodiesel. Bioresource Technology Reports. 2025;**29**:102018. DOI: 10.1016/j.biteb.2024.102018

[76] Sudalai S, Devanesan MG, Arumugam A. Dolomite as a potential source of heterogeneous catalyst for biodiesel production from *Pongamia pinnata*. Nature Environment and Pollution Technology. 2024;**23**(4). DOI: 10.46488/NEPT.2024.v23i04.042

[77] Suresh R, Ashwin R, Uppuluri K, MohanRaj T. Experimental investigation and operating characteristics of LHR engine using *Pongamia pinnata* and *Azadirachta indica* biodiesel blend with industrial waste additive. In: SAE Technical Paper 2025-28-0039. DOI: 10.4271/2025-28-0039

[78] Ali M, Kashif M, Zaidi AA, Jamil R. Statistical optimization of biodiesel production from non-edible *Pongamia pinnata* oil. Periodica Polytechnica Chemical Engineering. 2024;**68**(3):419-427. DOI: 10.3311/PPch.36753

[79] Kiani HS, Ali A, Zahra S, Hassan ZU, Kubra KT, Azam M, et al. Phytochemical composition and pharmacological potential of lemongrass (Cymbopogon) and impact on gut microbiota. AppliedChem. 2022;**2**(4):16. DOI: 10.3390/appliedchem2040016

[80] Gangolu NR, Gopidesi RK, Anudeep IM. Effects of lemongrass oil as fuel in diesel engine. INCAS

BULLETIN. 2020;**12**(4):63-70. DOI: 10.13111/2066-8201.2020.12.4.6

[81] Saravana Kumar M, Prabhahar M, Prakash S, Senthil J, Thiagarajan C. Experimental analysis of lemon grass biodiesel (LGB) with different exhaust gas recirculation. Materials Today: Proceedings. 2020;**33**(Pt. 1):876-880. DOI: 10.1016/j.matpr.2020.06.405

[82] Kumar P, Kumar D, Shankar R, Kumar S, Saini P, Kumar N. Effect of synthesized lemongrass biodiesel on the performance and emission characteristics of a CI engine. Sustainable Energy Technologies and Assessments. 2023;**57**:103221. DOI: 10.1016/j.seta.2023.103221

[83] Paramasivam P, Balasubramanian A, Adeleke AA, Ikubanni PP, Kumar S, Prakash C, et al. Exploring alternative fuel solutions: Lemon grass oil biodiesel blend with dibutyl ether additive for VCR diesel engines – An experimental analysis. Scientific Reports. 2024;**14**:20272

[84] Ooi JB, Kau CC, Manoharan DN, Wang X, Tran M-V, Hung YM. Effects of multi-walled carbon nanotubes on the combustion, performance, and emission characteristics of a single-cylinder diesel engine fueled with palm-oil biodiesel–diesel blend. Energy. 2023;**281**:128350. DOI: 10.1016/j.energy.2023.128350

[85] Ali MH, Mashud M, Rubel MR, Ahmad RH. Biodiesel from neem oil as an alternative fuel for diesel engine. Procedia Engineering. 2013;**56**:625-630. DOI: 10.1016/j.proeng.2013.03.169

[86] Dash SK, Elumalai PV, Ranjit PS, Das PK, Kumar R, Kunar S, et al. Experimental investigation on synthesis of biodiesel from non-edible neem seed oil: Production optimization and evaluation of fuel properties. Materials Today Proceedings. 2021;**47**(10):2463-2466. DOI: 10.1016/j.matpr.2021.04.551

[87] Fadairo AA, Wong PK, Ip WF, Ahmadi Ghadikolaei M, Cai Z, Ng KW, et al. Impact of neem oil biodiesel blends on physical and chemical properties of particulate matter emitted from diesel engines. Environmental Pollution. 2024;**362**:124972. DOI: 10.1016/j.envpol.2024.124972

[88] Arslan M, Ayyub H, Jamshaid M, Arslan A, Kalam MA, Ahmad F. Production, performance and emission of biodiesel from a mixture of castor oil and neem oil. RSC Advances. 2025;**15**(42):35296-35311. DOI: 10.1039/d5ra04004c

[89] Aswathanrayan MS, Santhosh N, Venkataramana SH, Kumar KS, Kamangar S, Arabi AIA, et al. Prediction of performance and emission features of diesel engine using alumina nanoparticles with neem oil biodiesel based on advanced ML algorithms. Scientific Reports. 2025;**15**:12683. DOI: 10.1038/s41598-025-97092-2

[90] Rajak AK, Harikrishna M, Zeenath SF, Dalal S, Karuna SL, Rajesh K, et al. Transesterification of neem seed oil for environmentally friendly biolubricants: Promoting circular economy in industrial processes. Biomass and Bioenergy. 2025;**200**:108012. DOI: 10.1016/j.biombioe.2025.108012

[91] Suzihaque MUH, Alwi H, Ibrahim UK, Abdullah S, Haron N. Biodiesel production from waste cooking oil: A brief review. Materials Today Proceedings. 2022;**63**(S1):S490-S495. DOI: 10.1016/j.matpr.2022.04.527

[92] Jarungwongsathien A, Chanthon N, Ngaosuwan K, Kiatkittipong W, Wongsawaeng D, Mens W, et al.

Cost-effective and simple biodiesel production process from waste cooking oil using a rotating tube reactor: Kinetic study and techno-economic analysis. Renewable Energy. 2025;**255**:123729. DOI: 10.1016/j.renene.2025.123729

[93] Alahmar NM, Wan Azelee NI, Toemen S. Efficient biodiesel production from waste cooking oil using a bifunctional Ce/Mn/$\gamma$-$Al_2O_3$ catalysts. Scientific Reports. 2025;**15**:352

[94] Rezaeifar A, Mansouri M, Maleki B. Incorporation of CuO on the $\alpha Fe_2O_3$ nanoparticles as a heterogeneous catalyst for conversion of waste cooking oil into biodiesel. Scientific Reports. 2025;**15**:7067

[95] Sahu S, Saikia K, Gurunathan B, Dhakshinamoorthy A, Rokhum SL. Green synthesis of CaO nanocatalyst using watermelon peels for biodiesel production. Molecular Catalysis. 2023;**547**:113342. DOI: 10.1016/j.mcat.2023.113342

[96] Selvam M, Pragadish N, Harish KA, Srinivasan RG, Devarajan Y, Kaliappan N. Investigating emission characteristics and combustion performance of a CRDI engine utilizing novel biodiesel derived from waste cooking oil and pentanol blends. Scientific Reports. 2025;**15**:14742

[97] Biberci MA, Çelik MB, Ozhuner E. Engine performance and emission optimization with waste cooking oil biodiesel/diesel blend using ANN and RSM techniques coupled with ACKTR-DE and HHO algorithms. BMC Chemistry. 2025;**19**(1):137. DOI: 10.1186/s13065-025-01512-3

[98] Liew OI, Razali NF, Mansur SA, Fauzi NA, Razak AHA, Abd Aziz MH, et al. Comparison of *Sapindus Mukorossi* (soapnut seed kernel oil) and used cooking oil via transesterification to produce biodiesel in terms of conversion percentage. ARAM. 2025;**136**(1):1220. DOI: 10.37934/aram.136.1.1220

[99] Bhran AA. A comparative techno-economic analysis of waste cooking oils and chlorella microalgae for sustainable biodiesel production. PRO. 2025;**13**(11):3526. DOI: 10.3390/pr13113526

[100] Dejene B, Bezu A. Biodiesel from blended microalgae and waste cooking oils: Optimization, characterization, and fuel quality studies. AIMS Energy. 2024;**12**(2):408-438. DOI: 10.3934/energy.2024019

[101] Aboueldahab M, Wu Y, Alwaleed E, Twagirayezu G, Xia A, Saber H. Maximizing the biodiesel production efficiency from Red Sea macroalgae: A seasonal and multivariate approach. Biomass and Bioenergy. 2025;**201**:108050. DOI: 10.1016/j.biombioe.2025.108050

[102] Pandian PM, Mathew F, Muttu M, Perumalsamy M. Techno-economic analysis of biodiesel production from *Chlorella vulgaris* using different potential biorefinery approaches. Bioenergy Research. 2025;**18**(1):21. DOI: 10.1007/s12155-025-10823-1

[103] Allasi HL, Murugan S, Ananda HK, Vairavel D. Enhancing diesel engine performance and emissions with N-butanol enhanced biodiesel derived from *Scenedesmus obliquus* algae. Scientific Reports. 2025;**15**:12924. DOI: 10.1038/s41598-025-92369-y

[104] Panwar P, Singh AP, Chauhan A, Arora R. Optimization of a macro-algae-based biodiesel supply chain: A multi-objective approach. Discover Sustainability. 2025;**6**:248. DOI: 10.1007/s43621-025-01022-1

[105] Liu Y, Xia R, Wang Y, Zhao Y, Ju M. Life-cycle environmental impact assessment of microalgae biodiesel derived from different processes. Industrial Crops and Products. 2025;**233**:121381. DOI: 10.1016/j.indcrop.2025.121381

[106] Sun H, Devanesan S, Aljawdah HM, Rithika M, Jhanani GK. Utilization of the *Dunaliella salina* microalgae by transesterification process with NaOH as catalyst in diesel combustion. Renewable Energy. 2025;**250**:123344. DOI: 10.1016/j.renene.2025.123344

[107] Poh ZL, Kadir WNA, Lam MK, Uemura Y, Suparmaniam U, Lim JW, et al. The effect of stress environment towards lipid accumulation in microalgae after harvesting. Renewable Energy. 2020;**157**:842-853. DOI: 10.1016/j.renene.2020.03.081

[108] Wu B, Zhou W, Gong Z, Ran S, He M, Zhong T, et al. Co-culture of *Chlorella pyrenoidosa* and *Cutaneotrichosporon* oleaginosum with enhanced biomass, lipid yield, and lipid productivity for micro-biodiesel production. Renewable Energy. 2025;**250**:123369. DOI: 10.1016/j.renene.2025.123369

[109] Al-Hamamre Z, Sandouqa A, Al-Saida B, Shawabkeh RA, Alnaief M. Biodiesel production from waste cooking oil using heterogeneous $KNO_3$/oil shale ash catalyst. Renewable Energy. 2023;**211**:470-483. DOI: 10.1016/j.renene.2023.05.025

[110] Karthikeyan M, Sathish S, Soosai MR, Prabu D, Venkatesan D. Optimized production of biodiesel from used (waste) cooking oil by utilizing a heterogeneous catalyst, and its studies for sustainable development. Fuel. 2026;**403**:136091. DOI: 10.1016/j.fuel.2025.136091

[111] Santoso A, Kusumah TN, Sumari S, Wijaya AR, Retnosari R, Rachman IB, et al. Synthesis of biodiesel from waste cooking oil using heterogeneous catalyst of $Na_2O/\gamma\text{-}Al_2O_3$ assisted by ultrasonic wave. AIMS Energy. 2022;**10**(5):1059-1073. DOI: 10.3934/energy.2022049

[112] Moumeni O, Djerad S, Bouranene S, Taghri M, Chakroun C. Production of biodiesel from waste cooking oils using manganese oxide MnO as a heterogeneous catalyst. Biofuels. DOI: 10.1080/17597269.2025.2544476

[113] Garg R, Garg R, Eddy NO, Abdulraheem MI, Hazzan OO, Abdi G. Green synthesis of calcium oxide nanocatalyst and application in transesterification of waste cooking oil. Bioresources and Bioprocessing. 2025;**12**:52

[114] Mishra R, Shu C-M, Gollakota ARK. Transformation of waste cooking oil into biodiesel in a microwave reactor using nickel-loaded graphitic biochar catalyst. Fuel. 2025;**397**:135369. DOI: 10.1016/j.fuel.2025.135369

[115] Helmiyati H, Budiman Y, Abbas GH, Dini FW, Khalil M. Highly efficient synthesis of biodiesel catalyzed by a cellulose@hematite-zirconia nanocomposite. Heliyon. 2021;**7**(3):e06622. DOI: 10.1016/j.heliyon.2021.e06622

[116] Gonzaga VE, Romero R, Gómez-Espinosa RM, Romero A, Martínez SL, Natividad R. Biodiesel production from waste cooking oil catalyzed by a bifunctional catalyst. ACS Omega. 2021;**6**(37):24092-24105. DOI: 10.1021/acsomega.1c03586

[117] Dhinagaran G, Vijayakumar G, Suvaitha SP, Harichandran G, Venkatachalam K. Conversion of neem oil (*Azadirachta indica*) to biodiesel

over SBA-15 supported sulphated zirconia catalysts. Catalysis Letters. 2024;**154**:2124-2139

[118] Hanif M, Bhatti IA, Hanif MA, Rashid U, Moser BR, Hanif A, et al. Nano-magnetic $CaO/Fe_2O_3$/feldspar catalysts for the production of biodiesel from waste oils. Catalysts. 2023;**13**(6):998. DOI: 10.3390/catal13060998

[119] Swarna SR, Srikanth HV, Anitha R, Swamy MT, Divakara TR, Krishnamurthy KN, et al. Biodiesel synthesis from mixed non-edible oil feedstock using $K_2CO_3$ in $NH_4OH$ catalyst mixture. Energy Sources, Part A: Recovery, Utilization and Environmental Effects. 2023;**45**(3):6901-6917. DOI: 10.1080/15567036.2023.2216163

[120] Saad M, Siyo B, Alrakkad H. Preparation and characterization of biodiesel from waste cooking oils using heterogeneous catalyst (Cat. TS-7) based on natural zeolite. Heliyon. 2023;**9**(6):e15836. DOI: 10.1016/j.heliyon.2023.e15836

[121] Foroutan R, Foroughi M, Tutunchi A, Ramavandi B, Terrón D, Pazos M. Production of biodiesel from waste cooking oil and non-edible Moringa oleifera oil utilizing a $CaO/CuFe_2O_4/LaTiO_3$ perovskite@Na bifunctional heterogeneous nanocatalyst. Journal of Environmental Chemical Engineering. 2025;**13**(4):117261. DOI: 10.1016/j.jece.2025.117261

[122] Perumal G, Magendradass D, Munusamy R, Gopal S. Optimization and kinetic studies on novel $Mg(HPW_{12}O_{40})/SiO_2$ mesoporous solid catalysts for biodiesel production from *Bauhinia variegata* seed oil. Chemical Engineering Research and Design. 2025;**218**:824-838. DOI: 10.1016/j.cherd.2025.05.029

[123] Hou S, Xie W. Exploring hierarchical porous $\gamma$-$Al_2O_3$ modified with MoV metal oxides used as a high-efficiency solid catalyst for sustainable biodiesel production from low-cost acidic oils. Renewable Energy. 2025;**250**:123303. DOI: 10.1016/j.renene.2025.123303

[124] Das S, Kaushik B, Chaudhury AP, Basumatary S, Pratap P, Mohan S, et al. Microwave-assisted biodiesel production from WCO using snail shell-derived CaO@coal fly ash: Optimization via RSM, cost analysis, kinetics, thermodynamics, and bibliometrics. Renewable Energy. 2025;**254**:123741. DOI: 10.1016/j.renene.2025.123741

[125] Wang Y, Hao X, Da X, Zheng X. Ultrasound-assisted dual-enzyme system for the preparation of fatty acid ethyl esters. Catalysis Letters. 2025;**155**:74

[126] Abulizi A, Bai H, Saierkejiang N, Okitsu K, Maihemuti M, Talifu D. Synthesis of biodiesel from waste cooking oil using a pulsed high-intensity ultrasound method. Energy Sources, Part A: Recovery, Utilization, and Environmental Effects. 2025;**47**(1):9435-9447. DOI: 10.1080/15567036.2021.1948935

[127] Yasvanthrajan N, Sivakumar P, Muthukumar K, Thanabalan M, Arunagiri A. Ultrasound assisted lipase catalysed transesterification using waste cottonseed oil. Journal of the Taiwan Institute of Chemical Engineers. 2025;**177**:106076. DOI: 10.1016/j.jtice.2025.106076

[128] Garuma WB, Bedru TK, Gindaba GT, Jayakumar M, Edae IT, Meshesha BT, et al. Optimization and characterization of ultrasonically extracted oil from watermelon seeds using response surface methodology: A pathway to biodiesel and epoxy oil

production. Biomass Conversion and Biorefinery. 2025;**15**:23135-23153

[129] Photaworn S, Tongurai C, Kungsanant S. Process development of two-step esterification plus catalyst solution recycling on waste vegetable oil possessing high free fatty acid. Chemical Engineering and Processing. 2017;**118**:1-8. DOI: 10.1016/j.cep.2017.04.013

Chapter 5

# The Use of Non-Food Biomass for Biodiesel Production: Economic Opportunities and Regulatory Framework [Part 2]

*Asma Nouira, Imen Amri, Imene Bekri-Abbes and Diogo Canavarro*

**Abstract**

The global energy transition requires a shift from fossil fuels to sustainable alternatives to mitigate climate change and enhance energy security. Biodiesel derived from non-food biomass has emerged as a pivotal solution for decarbonizing hard-to-abate sectors like transportation, overcoming the critical ethical and environmental limitations of first-generation biofuels. This chapter provides a comprehensive analysis of the non-food biodiesel sector, building on a foundation of advanced feedstocks and conversion technologies. It delves into the economic viability, policy frameworks, and socioeconomic dimensions essential for commercial scalability. The analysis highlights the significant opportunities in waste valorization and circular bioeconomy models, job creation in rural economies, and the crucial role of supportive policies and certifications. A comparative assessment of regulatory approaches in Europe, the United States, and Asia is presented, alongside case studies of successful initiatives. The chapter concludes with future perspectives on technological innovation, the role of biodiesel in decarbonizing aviation and maritime transport, and synergies with other renewables. Finally, it offers strategic recommendations for policymakers and industry stakeholders to foster a sustainable, equitable, and integrated biodiesel industry.

**Keywords:** biodiesel, non-food biomass, circular economy, policy incentives, sustainability, advanced biofuels, decarbonization, renewable energy

## 1. Introduction

The global energy sector stands at a critical juncture, driven by the dual imperatives of climate change mitigation and enhanced energy security. For over a century, the world's economic and industrial progress has been overwhelmingly powered by fossil fuels, a dependency that has led to profound environmental degradation, geopolitical instability, and resource challenges. In response, a global transition toward sustainable, renewable energy sources is not merely an option but an urgent necessity [1]. Within this broader transformation, biofuels have emerged as a cornerstone of the strategy to decarbonize hard-to-abate sectors, most notably transportation. Among these, biodiesel, a renewable alternative to conventional diesel fuel, has gained considerable traction due to its potential to significantly reduce greenhouse gas emissions and leverage existing fuel infrastructure [2]. Historically, the development of biofuels has progressed through distinct generations, each seeking to overcome the limitations of its predecessor. First-generation biofuels, derived from edible crops like corn, palm oil, and sugarcane, provided a proof of concept but also ignited widespread public and scientific debate. The "food versus fuel" dilemma underscored significant ethical and environmental concerns, as large-scale cultivation of energy crops competed with land use for food production, contributing to food price volatility and driving deforestation in many regions [2]. The realization that the carbon savings from these fuels were often negated by emissions from agricultural practices and land-use change catalyzed a critical re-evaluation of the entire biofuel paradigm. This pivotal shift has propelled the industry toward a more sustainable and ethical pathway, focusing on second- and third-generation biofuels derived from non-food biomass [3]. This chapter subdivided into two parts comprehensively explores this new frontier, building on a foundational understanding of the feedstocks and conversion technologies. The first part of chapter has provided a detailed typology of these advanced feedstocks, highlighting their clear environmental and social advantages. It explored the immense potential of agricultural and forestry residues (such as rice husk, sugarcane bagasse, and wood chips) that offer a low-cost supply with minimal land-use impact. Additionally, it examined the promise of non-edible energy crops (like *Jatropha curcas* and *Pongamiapinnata*) that can thrive on marginal land, thereby avoiding competition with food systems. The chapter also delved into the valorization of organic waste (including used cooking oil and animal fats) as a key component of a circular bioeconomy model, and the innovative use of micro- and macroalgae, which boast exceptionally high oil content and can sequester carbon dioxide. Furthermore, the first part of this chapter also provided an in-depth analysis of the technological landscape for converting these diverse feedstocks into biodiesel. We have discussed the evolution from conventional homogeneous catalysis, to more advanced and sustainable methods. Building on this critical foundation, the second part of this review chapter delves into the economic, policy, and social dimensions that are essential for the widespread commercialization and long-term success of the non-food biodiesel industry. This section will begin by analyzing the key economic opportunities, focusing on waste valorization and the principles of a circular bioeconomy as a framework for creating new revenue streams and fostering rural development. We will then critically examine the crucial role of regulatory frameworks and policy incentives, including international standards, sustainability certifications, subsidies, and carbon credit systems, providing a comparative analysis of public policies across Europe, the United States, and Asia. We will also present case studies of successful initiatives to distill key lessons learned and best practices. Ultimately, this review aims to provide a holistic and forward-looking perspective on the multifaceted challenges and opportunities facing the non-food biodiesel industry, charting a clear and sustainable path towards its full integration into a global energy system that is both equitable and environmentally responsible.

## 2. Economic opportunities and business models for biodiesel

### 2.1 Waste valorization and circular economy

Economic opportunities in the biodiesel industry are significantly enhanced through waste valorization and the adoption of circular economy principles [1]. This approach not only provides a sustainable alternative to fossil fuels but also addresses environmental challenges by converting waste into valuable resources [2]. A key aspect of this business model involves utilizing diverse waste sources, such as used cooking oil, animal fats, and agricultural residues, which often have low or even negative acquisition costs. For example, the valorization of defatted cottonseed into biochar, which then acts as a catalyst for biodiesel production, demonstrates a closed-loop system that enhances the economic viability of the cotton industry and can potentially replace fossil fuel requirements for cotton cultivation [2]. Similarly, the recycling of used cooking oil for biodiesel production, along with other value-added products like waste plastics, integrates waste management with resource recovery, aligning with circular economy practices [3]. Furthermore, advancements in catalyst reuse, such as the application of heterogeneous ion-exchange resin catalysts for esterification, significantly lower the environmental footprint and offer substantial cost efficiencies, leading to quick payback periods and high returns-on investment [4]. Innovative approaches like using deep eutectic solvents (DES) for biodiesel synthesis from greasy wastewater sludge highlight the potential for greener solutions and considerable reductions in greenhouse gas emissions through efficient waste remediation [5]. The concept of a circular bioeconomy is further strengthened by integrating carbon capture and waste remediation using micro-algae, which can harness both inorganic $CO_2$ and organic agro-wastes to produce sustainable biodiesel, offering triple benefits of carbon sequestration, waste treatment, and biofuel production [6]. Even waste agricultural by-products like rice husk ash and orange peel can be transformed into effective heterogeneous catalysts for biodiesel production, showcasing the excellent recyclability and addressing waste disposal issues [7]. These examples underscore how waste valorization and circular economy models are not just environmentally beneficial but also create significant economic opportunities, making biodiesel production more sustainable and competitive [8, 9].

### 2.2 Job creation and rural development

The growth of the biodiesel industry presents significant opportunities for job creation and rural development, particularly in regions undergoing sustainable energy transitions [10]. By leveraging locally available biomass and waste feedstocks, biodiesel production can stimulate rural economies, generate additional income streams for smallholder farmers, and create the employment opportunities across the entire value chain. This includes activities such as feedstock cultivation, collection, processing, transportation, and distribution [11, 12]. Such a decentralized approach to energy production not only empowers rural communities but also reduces dependence on imported fossil fuels, enhancing energy security [10]. The optimization of supply chains within the biodiesel sector further amplifies these benefits. Improved transparency, strengthened quality control, and streamlined logistics reduce production losses, ensure consistent fuel quality, and support sustainable resource management. These enhancements translate into increased employment opportunities in rural areas, as skilled labor is required for monitoring, quality assurance, and efficient transportation [13]. Additionally, adopting circular economy practices—such as reusing by-products and integrating waste streams—fosters sustainable agricultural development while reinforcing local job markets. Policy frameworks play a crucial role in maximizing these socioeconomic benefits. National strategies, such as Ghana's Bioenergy Policy, explicitly recognize job creation and economic growth as the essential outcomes of biofuel integration [14]. Moreover, the expansion of bio-based industries in the Global South is increasingly seen as a vital driver for sustainable economic development, enabling communities to capitalize on renewable biological resources, diversify local economies, and improve livelihoods [15]. Overall, the biodiesel sector represents a multifaceted opportunity to stimulate rural development, enhance employment, and contribute to broader sustainability objectives.

### 2.3 Integration into national and international energy policies

The integration of biodiesel into national and international energy policies is a critical step toward achieving sustainable development goals, enhancing energy security, and mitigating climate change (**Table 1**) [14]. Governments worldwide are increasingly recognizing biodiesel's potential and are implementing diverse policy measures to support its production and adoption [14, 15]. These policies often include blending mandates, fiscal incentives like tax subsidies, and strategic investments in infrastructure [14–16].

For instance, India has initiated various policies, such as the National Policy on Biofuels and an ethanol blending program, to reduce its reliance on fossil fuels and meet ambitious blending targets [16]. Similarly, Indonesia's biofuel policy, centered on palm oil-based biodiesel, has seen blending reach significant levels (B35), with further efforts to enhance palm oil productivity through replanting schemes that also aim to reduce poverty and boost regional economies [17]. Brazil, a major biofuel producer, is strategically positioning biofuels as drivers for decarbonization and economic transformation, with policies that favor advanced biofuels and align with no-regret approaches to ensure food security [18]. The European transport sector

| Country/region or policy area | Policy strategies | Primary objectives |
|---|---|---|
| India | National Policy on Biofuels, ethanol blending mandates, fossil fuel reduction targets. | Reduce fossil fuel reliance, meet blending targets. |
| Indonesia | Palm oil-based biodiesel blending (B35), replanting schemes to boost productivity and regional development. | Enhance energy security, reduce poverty, promote economic growth. |
| Brazil | Support for advanced biofuels, alignment with food security, economic decarbonization strategy. | Decarbonize economy, ensure sustainable energy and food systems. |
| European Union | Carbon pricing, GHG intensity targets, mandates for renewable fuel share in transport. | Create low-carbon fuel supply chains, enhance sustainability in transport. |

**Table 1.**
*Policy strategies supporting biodiesel integration.*

is also exploring various policy measures, including carbon pricing, GHG intensity reduction, and renewable fuel share mandates, to establish robust renewable fuel supply chains [19]. Beyond national policies, the concept of a circular economy is influencing urban planning and port diversification strategies, transforming fossil fuel hubs into circular economy centers that integrate biowaste supply with biofuel demand [20]. These policies collectively aim to create a benign market environment, optimize supply chains, and reduce the carbon footprint of biodiesel, demonstrating a global commitment to a more sustainable energy future [21–23].

### 2.4 Case studies: Successful initiatives and lessons learned

Successful initiatives and valuable lessons learned from various case studies highlight the diverse pathways to establishing a viable biodiesel industry. In Indonesia, an optimization study for the palm oil-sourced biodiesel supply chain focused on minimizing costs by strategically determining the locations of blending terminals and new biodiesel plants, demonstrating the importance of detailed supply chain modeling for national mandates [24]. Consumer satisfaction in Southern Thailand indicated high acceptance of biodiesel products and pricing, underscoring the significance of product availability and effective promotion in market penetration [25]. Iran has explored complex, multimodal, and multi-period biodiesel supply chain networks, integrating various feedstocks like Jatropha, waste cooking oil, and microalgae, proving the applicability of mixed-integer linear programming models for optimizing design under uncertainty [26]. Further case studies reveal the critical technical and operational insights. Research on modeling biodiesel properties, such as cetane number from fatty acid methyl esters, helps in predicting performance and optimizing composition, reducing the need for extensive experimental trials [27]. Economic assessments, as seen in Brazil, have analyzed the production costs of biodiesel from different feedstocks like soybean, sunflower, chicken oil, and beef tallow, finding that beef tallow offered the lowest production costs in their experimental plant [28]. Efforts in waste valorization, such as using waste carbon sources and blended non-edible/waste oils with carbon-based catalysts, demonstrate how feedstock flexibility and cost-effective catalysts can enhance the industry's economic viability [29]. Life cycle assessments, including a study on Jatropha-based biodiesel in Thailand and soybean/algae biodiesel for US marine vessels, emphasize the need to evaluate the environmental and social impacts comprehensively, from cultivation to end-use, to ensure true sustainability and inform policy [30, 31]. In Mexico, analysis of the castor oil plant supply chain highlighted the feasibility of establishing biofuel supply chains by organizing farmers and managing the process from cultivation to sale [32]. Furthermore, studies in China on optimizing waste cooking oil for biodiesel supply chains, integrating economic and environmental objectives with GIS modeling, show how advanced frameworks can manage location, production, inventory, and distribution decisions [33]. Lessons from South Africa emphasize the importance of occupational health and safety in small-scale biodiesel plants, pointing to resource limitations for compliance [34]. Brazilian case studies also demonstrate how institutional arrangements, particularly farmer collective action and technical/financial support, are crucial for integrating small family farmers into the emerging biodiesel industry [35]. Finally, Fiat's plant in Brazil successfully implemented a case study on producing biodiesel from frying oil for their employee transport minibuses, showcasing a practical application of waste-to-energy conversion and assessing its environmental viability through a life cycle assessment [36].

## 3. Regulatory framework and policy incentives

### 3.1 Standards and certifications for biodiesel from non-food biomass

For biodiesel to achieve widespread commercialization and market acceptance, it must adhere to stringent quality assurance standards and certifications [37, 38]. These standards ensure the fuel's performance, protect engine durability, and verify its environmental benefits [37]. Key international standards-setting bodies include the American Society for Testing and Materials (ASTM), the European Committee for Standardization (CEN), and the International Organization for Standardization (ISO) [38]. For instance, ASTM D6751 is the primary standard for biodiesel fuel blend stock (B100) in the United States, while EN 14214:2002 (and its subsequent revisions) is the European standard [39–43]. These standards specify various properties, including flash point, water and sediment content, viscosity, sulfated ash, sulfur content, corrosion, cetane number, acid number, carbon residue, and most importantly, free and total glycerin content [39, 40, 42]. Beyond product specifications, sustainability certifications play a crucial role in validating environmentally and socially responsible production practices. The Roundtable on Sustainable Biomaterials (RSB) certification, for example, has been earned by leading producers of waste-based biodiesel, signifying adherence to sustainability criteria throughout the production process [44]. Similarly, the Indonesian Sustainable Palm Oil (ISPO) certification system aims to ensure the sustainability of palm oil production, which is a major feedstock for biodiesel in Indonesia [45]. The evolving nature of these standards necessitates continuous review and improvement, as seen in China's efforts to establish and refine its biodiesel standard system band in Indonesia's validation of monoglyceride and water content for B30 implementation [46, 47]. The challenges of maintaining biodiesel purity, especially with varying feedstocks and production processes, underscore the importance of robust analytical methodologies and calibration standards to ensure compliance with these regulations [48]. The ongoing development and adherence to these standards and certifications are vital for fostering trust in biodiesel as a sustainable and reliable alternative fuel.

### 3.2 Subsidies, carbon credits, and support mechanisms

Subsidies, carbon credits, and various support mechanisms play a crucial role in enhancing the economic viability and promoting the adoption of biodiesel, particularly in its early stages of development and expansion [49]. Governments globally are implementing policies that recognize the environmental benefits of biodiesel, such as reduced greenhouse gas emissions, by offering incentives that bridge the cost gap with fossil fuels [49–51]. Carbon credits, often generated through mechanisms like Brazil's National Biofuels Policy (RenovaBio), provide a significant financial incentive by allowing biodiesel producers to sell "decarbonization credits" to fossil fuel distributors or other interested parties [50–52]. This system quantifies the environmental benefit of biofuels, translating GHG emission reductions into a marketable commodity [50–52]. For instance, a Brazilian case study estimated that mandatory biodiesel blending could prevent the emission of at least 170 million tons of carbon dioxide equivalent between 2023 and 2030, with an average carbon intensity significantly lower than pure diesel [50]. Such credits can substantially reduce the operating costs of biodiesel production, making the process more economically attractive [49, 51, 52]. Beyond carbon credits, direct subsidies and other financial support mechanisms are essential. These can include grants for plant construction, tax exemptions, and guaranteed purchase prices or blending mandates that create a stable market demand for biodiesel [50–52]. For example, the integration of microalgae biodiesel production with sugarcane biorefineries in Brazil, while currently less profitable than traditional ethanol production, becomes economically viable when carbon credits are considered [51]. Similarly, analyses in Brazil demonstrate how the sale of excess biodiesel, glycerin, and carbon credits from waste frying oil conversion can generate significant profits and contribute to urban sustainability [52]. These support mechanisms are vital for de-risking investments in biodiesel infrastructure and fostering innovation in feedstock utilization and production processes [53].

### 3.3 Comparison of public policies: Europe, United States, and Asia

Public policies regarding biodiesel exhibit significant variations across continents, reflecting diverse energy priorities, agricultural landscapes, and environmental goals (**Table 2**). In Europe, early and robust support for biodiesel development has made it the world's largest production and consumption region [54, 55, 72]. European policies have emphasized reducing greenhouse gas emissions and decreasing reliance on fossil fuels, often through blending mandates and tax incentives [55–58]. The focus has largely been on first-generation feedstocks like rapeseed (winter rape) in Southern Europe, though concerns about nitrous oxide emissions from fertilizer use and the sustainability of crop-based biofuels have led to a shift toward waste-based feedstocks, such as used cooking oil [56, 57]. The European Union has also established a sophisticated and continuously evolving standard system (e.g., EN 14214) for biodiesel products, playing a crucial role in regulating and safeguarding the industry's development and influencing standards in other regions like China [58]. In the United States, biodiesel policy approaches have varied at both federal and state levels. Federal policies, such as the Renewable Fuel Standard, have driven demand through renewable identification number (RIN) credits, while state-level policies have focused on infrastructure development and production support [59–61]. The US market has seen a significant emphasis on soybean oil-based biodiesel, with projections of substantial demand from sectors like federal fleets, mining, and marine areas [62]. However, the efficacy of policies, particularly concerning the interconnectedness of fuel and food markets, is a continuous area of study [63]. Quality standards, primarily ASTM D6751, are critical for ensuring fuel performance and engine compatibility, with ongoing efforts to monitor and improve the quality of biodiesel blends [64–66]. Challenges remain in expanding pipeline transport for biodiesel and fully understanding its impact on NOx emissions [66]. Asia, particularly Southeast Asia, is emerging as a major player in global biodiesel production, largely driven by its abundant palm oil resources [67, 68]. Countries like Indonesia and Malaysia have invested heavily in developing their biodiesel industries, with policies centered on increasing domestic blending mandates and promoting sustainable palm oil certification [67–69]. While palm oil offers significant production potential, its environmental impacts, such as deforestation for new plantations, are growing concerns [68, 69]. Efforts are also underway in East Asia to harmonize biodiesel fuel standards to facilitate regional trade and maximize benefits from production and utilization, although practical implementation can face political hurdles [70]. Furthermore, research into alternative feedstocks like rubber seed oil demonstrates a drive to diversify production and enhance regional energy security [71].

| Feature | Europe (EU) | United States (US) | Asia (primarily Southeast Asia) |
|---|---|---|---|
| Overall stance/ focus | Early and robust support; world's largest producer/consumer. Emphasis on GHG reduction and fossil fuel reliance decrease. | Varied approaches (federal and state). Emphasis on demand creation (RINs) and production support. | Emerging major player, driven by abundant palm oil. Focus on domestic blending and sustainable certification. |
| Key policies/ mechanisms | Blending mandates, tax incentives. Sophisticated and evolving standard system (e.g., EN 14214). | Renewable Fuel Standard (federal) with RIN credits. State-level infrastructure development and production support. | Increasing domestic blending mandates. Promotion of sustainable palm oil certification. |
| Primary feedstocks | Historically rapeseed (winter rape); shifting toward waste-based feedstocks (e.g., used cooking oil). | Predominantly soy bean oil. | Primarily palm oil. Research into alternative feedstocks (e.g., rubber seed oil) for diversification. |
| Environmental/ sustainability concerns | Nitrous oxide emissions from fertilizer (rapeseed); sustainability of crop-based biofuels. Shift to waste-based due to these concerns. | Interconnectedness of fuel and food markets (continuous study). Impact on NOx emissions (ongoing challenge). | Environmental impacts of palm oil (e.g., deforestation for new plantations). |
| Quality standards | Continuously evolving EN 14214 (influences other regions). | ASTM D6751. Ongoing efforts to monitor and improve quality. | Efforts to harmonize biodiesel fuel standards for regional trade. |
| Challenges/notes | Shifting from crop-based to waste-based feedstocks. | Expanding pipeline transport; fully understanding NOx impact. | Environmental impacts of palm oil cultivation; practical implementation hurdles for harmonized standards. |
| References cited | [54–58] | [59–66] | [67–71] |

**Table 2.**
*Comparison of public policies: Europe, United States, and Asia.*

## 4. Future perspectives and recommendations

### 4.1 Technological innovations toward 2030–2050

Technological innovations are rapidly shaping the future of biodiesel production, with a clear trajectory toward more sustainable, efficient, and cost-effective methods for the period 2030–2050 [73]. A primary focus is on diversifying feedstocks beyond traditional edible oils to address concerns about food security and land use. Industrial sludge, for instance, is emerging as a viable and readily available feedstock, offering significant potential for waste reduction and greenhouse gas emission mitigation, though scalability remains a challenge [74, 75]. Similarly, the potential of fat, oil, and grease (FOG) from various waste streams is being critically reviewed, with advancements needed in efficient conversion methods for high free fatty acid content [76]. Non-edible oils, such as those from *Camelina sativa* and various other plant sources, are gaining prominence due to their high yield, low production cost, and reduced competition with food crops [77, 78]. Microbial oils, derived from microalgae, fungi, and bacteria, are also seen as a promising third-generation feedstock, offering advantages such as a small footprint, high lipid content, and efficient $CO_2$ uptake, despite current challenges in cultivation, harvesting, and extraction for cost-effective production [79, 80].

Beyond feedstock diversification, significant advancements are expected in conversion technologies and catalytic processes. Supercritical transesterification, a catalyst-free method, is being explored for its faster reaction rates, high conversion efficiency, and environmental friendliness, with ongoing research into optimizing parameters and integrating energy recycling [81]. The development of novel catalysts, particularly heterogeneous catalysts like calcium oxide (CaO)-based materials, is crucial for improving reaction efficiency, reusability, and reducing purification costs [82, 83]. Additionally, advancements in biorefinery concepts are integrating biodiesel production with other valuable co-products, enhancing overall economic viability. The role of advanced reactor technologies, such as plug-flow and microwave reactors, is also being comprehensively reviewed to identify the optimal designs for energy and time efficiency in large-scale biodiesel synthesis [84]. Finally, the broader energy transition, particularly in countries like Brazil, is seeing renewable diesel emerge as a complementary biofuel to traditional biodiesel, offering the enhanced properties and further contributing to environmental goals within existing biofuel policies [85]. These innovations collectively aim to overcome current economic and technical hurdles, making biodiesel a more competitive and sustainable alternative to fossil fuels in the coming decades [86, 87].

### 4.2 Potential role in decarbonizing hard-to-electrify sectors (aviation and maritime transport)

Biodiesel is increasingly recognized for its crucial role in decarbonizing hard-to-electrify sectors like aviation and maritime transport, where direct electrification remains challenging [87, 88]. In the maritime transport sector, biodiesel, particularly derived from waste vegetable oils and animal fats, is emerging as a viable short-term solution for immediate emission reductions [87, 89]. Studies have confirmed biodiesel's potential to significantly reduce $CO_2$ emissions (by about 30%) and CO emissions (up to 94%) in marine diesel engines, while maintaining engine performance, though a trade-off of increased NOx emissions (around 30%) is a known factor [89]. The International Maritime Organization (IMO) has highlighted the importance of alternative marine fuels like biodiesel and renewable diesel as "drop-in fuels" that can be used with existing engine systems with minimal or no modifications, helping to meet the initial greenhouse gas (GHG) reduction targets [90]. From a refinery perspective, pathways to produce very low sulfur fuel oil (VLSFO) for marine applications include direct blending of biodiesel and renewable diesel, as well as co-processing pyrolysis oils, with bio-VLSFO production often proving cost-effective compared to fossil VLSFO at crude oil prices [91]. This cost-effectiveness, combined with the relatively relaxed specifications for marine fuels and the low sulfur content of biofuels, positions marine fuels as a favorable entry point for refiners to begin decarbonizing with biofuels [91]. For the aviation sector, sustainable aviation fuels (SAF), which include biojet fuel derived from waste feedstocks such as used cooking oils and lignocellulosic materials, represent the most viable solution to address the climate change and meet growing fuel demand [87]. While first- and third-generation biofuels face challenges, second-generation biodiesel (e.g., from used cooking oil) demonstrates a significantly lower carbon and environmental footprint compared to petrodiesel, making it a promising stepping-stone for decarbonizing the transport sector [88]. Furthermore, advancements in biofuel blends, such as hydrotreated used cooking oil (HVO) blended with biodiesel and diesel, show the promising results in enhancing thermal efficiency and reducing emissions of hydrocarbons (HC), carbon monoxide (CO), and smoke in compression ignition engines [92]. The integration of gasoline into biodiesel-diesel mixtures is also being explored as a solution to reduce GHG emissions in engines, by improving combustion characteristics and suppressing soot, CO, and unburned hydrocarbon emissions, although favorable legislation and incentives are needed for wider deployment [93]. Even the rail sector, despite being largely electrified, still relies on oil for a significant portion of its energy consumption, making biodiesel a high-potential alternative fuel for reducing carbon footprint in locomotive engines through the adoption of drop-in fuels [94]. The continuous research and development efforts across various transportation modes underscore the biodiesel's increasing importance as a multi-sector decarbonization strategy.

### 4.3 Synergies with other renewable energy sources

Biodiesel, as a versatile biofuel, offers significant opportunities for synergy with other renewable energy sources, contributing to a more robust and diversified sustainable energy system. These synergies can manifest in various ways, from utilizing renewable energy to power biodiesel production to using biodiesel as a flexible energy source in hybrid systems. One key synergy lies in powering biodiesel production with renewable energy. The transesterification process, a common method for biodiesel synthesis, requires energy inputs, particularly for heating. Integrating solar thermal energy, for instance, through parabolic trough collectors, can directly supply the heat needed for the reaction, significantly reducing the carbon footprint of the production process and enhancing overall energy efficiency [95]. This creates a truly "green" biodiesel, as the entire lifecycle, from feedstock to fuel, is powered by renewable sources. Biodiesel also plays a vital role in hybrid energy systems, particularly for managing the intermittency of other renewables such as solar and wind. While solar and wind are excellent for electricity generation, their output fluctuates depending on weather conditions. Biodiesel-powered generators can provide reliable, on-demand power to balance these fluctuations, acting as a form of energy storage that can be dispatched when needed [96]. Beyond direct power generation, biodiesel can be the strategic component in waste-to-energy pathways. The utilization of waste feedstocks, such as used cooking oil or industrial sludge, for biodiesel production not only provides a renewable fuel but also addresses the waste management challenges. This aligns with circular economy principles, maximizing resource utilization and minimizing environmental impact. The by-products of biodiesel production, like glycerin, can even be utilized in other biomass conversion processes, such as anaerobic digestion for biogas production, creating a "hybrid biorefinery" concept that leverages the economic and energy synergies. Finally, the inherent characteristics of biodiesel, such as its liquid form and relatively high energy density, make it a valuable energy carrier and storage medium. Unlike electricity from solar or wind, which requires expensive battery storage for large-scale, long-duration applications, biodiesel can be stored and transported relatively easily. This makes it particularly suitable for sectors where liquid fuels are essential, such as heavy-duty transport (trucking, rail, maritime) and aviation, where direct electrification is technically challenging or economically prohibitive in the short to medium term.

### 4.4 Recommendations for policymakers and industry stakeholders

#### *4.4.1 Develop clear and consistent long-term policy frameworks*

Policymakers are urged to establish clear and consistent long-term policy frameworks to foster the sustainable growth of the biodiesel industry. This includes implementing ambitious yet realistic blending mandates, such as B20, B30, or B100 where feasible, with defined long-term targets stretching to 2030, 2040, and 2050, thereby providing the predictability necessary for industry to confidently invest in the production capacity and infrastructure. Furthermore, policies should differentiate and specifically support advanced biofuels derived from waste feedstocks such as used cooking oil, industrial sludge, and agricultural residues, as well as non-edible crops, to prevent the competition with food production and mitigate indirect land-use change impacts [76, 78]. It is also crucial to promote the adoption of internationally recognized quality standards, such as ASTM D6751 and EN 14214, alongside sustainability certifications such as RSB and ISPO, which are essential for ensuring product quality, facilitating international trade, and verifying environmental and social benefits [37, 38, 41]. To bridge the cost gap with fossil fuels and stimulate investment, policymakers should strengthen carbon pricing mechanisms and carbon credit schemes, drawing inspiration from models like Brazil's RenovaBio, which directly reward greenhouse gas emission reductions achieved by biodiesel [50]. Finally, offering targeted tax exemptions, grants, and subsidies for biodiesel production, especially for small- and medium-scale enterprises and those employing innovative feedstocks or technologies, will de-risk investments and foster innovation across the sector [49].

#### *4.4.2 Invest in research, development, and demonstration (RD&D)*

To achieve sustainable biodiesel production, it is essential to invest in Research, Development, and Demonstration (RD&D). First, supporting novel feedstock development through funding for high-yield, sustainable, and non-food-competing sources—such as microalgae, Jatropha, Camelina, and waste streams—can reduce land-use conflicts and enhance feedstock availability [33, 77, 79]. Second, advancing conversion technologies *via* R&D in supercritical transesterification, enzymatic processes, and catalytic improvements can increase efficiency and reduce costs, making biodiesel more competitive [81–83].

#### *4.4.3 Facilitate infrastructure development*

To enable the widespread adoption of biodiesel, infrastructure development must be prioritized. First, investments should be made in blending, storage, and distribution infrastructure, including upgrades to existing facilities and the expansion of dedicated biodiesel pipelines where feasible, ensuring seamless integration into fuel supply chains. Second, for hard-to-electrify sectors such as aviation and maritime, policies should support the development of refueling infrastructure for sustainable aviation fuels (SAF) while recognizing biodiesel's role as a transitional or complementary fuel [87, 88]. By strengthening both conventional and specialized infrastructures, biodiesel can be efficiently scaled, meeting demand across transportation sectors and facilitating a smoother transition to low-carbon alternatives.

#### *4.4.4 Promote public awareness and education*

To accelerate biodiesel adoption, the targeted efforts in public awareness and workforce development are crucial. First, clear communication campaigns should highlight the biodiesel's environmental and economic benefits—such as reduced greenhouse gas emissions, energy security, and rural job creation [25]. Engaging consumers, fleet operators, and policymakers can build confidence and drive demand. Second, workforce training programs must be established to cultivate expertise across the biodiesel value chain, including feedstock production, advanced conversion technologies, and distribution logistics. By investing in education and skills development, the industry can ensure a capable workforce to support scaling efforts while fostering long-term economic growth. Together, these initiatives will strengthen market acceptance and industry readiness for biodiesel expansion.

#### *4.4.5 Prioritize sustainable feedstock sourcing*

To ensure the long-term viability of biodiesel, sustainable feedstock strategies must be a top priority. First, waste-to-biodiesel production should be expanded by maximizing the use of waste cooking oil, animal fats, and other waste lipid streams, reducing landfill burdens while enhancing circularity in fuel production [76]. Second, non-edible oilseed cultivation—such as Jatropha and Camelina—should be promoted through investments in agronomic research and marginal land utilization, avoiding competition with food crops [78]. Finally, robust traceability and certification systems must be implemented to verify sustainable sourcing, prevent deforestation-linked feedstocks, and strengthen consumer and regulatory confidence [44, 45]. By prioritizing these measures, the biodiesel industry can minimize its environmental footprint while securing a scalable, ethically sourced supply chain.

### 4.5 Optimize production processes for efficiency and cost-effectiveness

To strengthen biodiesel's competitiveness, continuous improvements in production efficiency and cost reduction are essential. First, advanced catalytic technologies—particularly reusable heterogeneous catalysts—should be prioritized to enhance reaction efficiency, lower costs, and minimize waste generation [82, 83]. Second, adopting integrated biorefinery models can maximize economic returns by co-producing high-value by-products (e.g., purified glycerin for pharmaceuticals or cosmetics), turning waste streams into revenue streams [29]. Finally, energy-efficient production practices, such as heat recovery systems and renewable energy integration (e.g., solar thermal), can significantly reduce both operational expenses and the carbon footprint of biodiesel facilities. By refining these processes, the industry can achieve greater scalability, sustainability, and economic resilience.

### 4.6 Improve catalytic technologies

To enhance the sustainability and economic viability of biodiesel production, a multi-faceted approach to process optimization is essential. First, significant advancements in catalytic technologies, particularly the development of robust, reusable heterogeneous catalysts, can dramatically improve reaction efficiency while reducing both production costs and environmental impact [82, 83]. Second, the implementation of integrated biorefinery models presents an opportunity to maximize resource utilization by co-producing high-value by-products such as pharmaceutical-grade glycerin from biodiesel by-products, thereby improving overall process economics [29]. Third, prioritizing energy efficiency through the adoption of advanced process engineering and renewable energy integration—such as solar thermal systems for heating requirements—can significantly reduce both operational costs and the carbon footprint of production facilities [1, 2]. These synergistic improvements in catalysis, biorefinery concepts, and energy management collectively contribute to a more competitive and sustainable biodiesel industry, better positioned to meet both economic and environmental objectives while adhering to circular economy principles.

## 5. Invest in product quality and diversification

To strengthen biodiesel's market position and versatility, strategic investments in product quality and diversification are critical. First, the industry must exceed existing quality standards (such as ASTM D6751 and EN 14214) through rigorous testing and process refinement, ensuring seamless engine compatibility while minimizing maintenance concerns [38, 39, 65]. Second, application-specific fuel formulations should be developed—including cold-weather additives or marine/aviation-optimized blends—to address niche markets and expand biodiesel's usability across sectors [55, 89]. By combining stringent quality control with tailored fuel solutions, biodiesel can enhance its reliability, consumer trust, and adaptability to diverse the operational demands. These efforts will be particularly vital for competing with conventional fuels and supporting decarbonization in hard-to-electrify transport sectors.

### 5.1 Meet and exceed quality standards

To solidify biodiesel's role in the energy transition, a dual focus on quality excellence and application-specific innovation is paramount. First, the industry must not only meet but surpass global quality benchmarks (ASTM D6751, EN 14214) through advanced refining processes and stringent testing protocols. This ensures superior engine performance, reduces wear, and builds consumer confidence—particularly crucial for gaining OEM approvals and competing with petroleum diesel [38, 39]. Second, the targeted blend development can unlock new markets: Cold-weather formulations with improved low-temperature flow properties, or marine/aviation-optimized blends with enhanced stability, could decarbonize sectors where electrification remains challenging [55, 58]. These efforts should be coupled with collaborative R&D between fuel producers, engine manufacturers, and policymakers to validate the performance and accelerate adoption. By marrying uncompromising quality with tailored solutions, biodiesel can transition from a generic alternative to a precision fuel capable of meeting diverse sectoral needs while supporting the broader emission reduction goals.

## 6. Foster collaboration and partnerships

To accelerate the sustainable growth of biodiesel, fostering robust partnerships across sectors and borders is essential. The industry should prioritize the cross-sector alliances that connect agricultural innovators, waste management specialists, engine designers, and transportation leaders to create a seamless, optimized value chain—from feedstock sourcing to end-use applications. Concurrently, global knowledge-sharing initiatives can propel the sector forward, enabling countries to exchange the technical breakthroughs, policy frameworks, and market intelligence that accelerate worldwide adoption. These collaborative approaches not only strengthen domestic biodiesel ecosystems but also position renewable fuels as the key players in international climate solutions. When combined with the previously outlined strategies in feedstock development, production optimization, and quality enhancement, such partnerships create a powerful framework for transforming biodiesel into a cornerstone of the low-carbon energy transition—delivering measurable benefits for energy security, emission reduction, and circular economic principles across local and global scales. By implementing these recommendations, policymakers and industry stakeholders can collectively drive the sustainable growth of the biodiesel sector, contributing significantly to energy security, climate change mitigation, and a more circular economy.

## 7. Summary of key points

The global transition toward renewable energy has positioned biodiesel as a critical solution for reducing greenhouse gas emissions, enhancing energy security, and promoting sustainable development. Derived from non-food biomass—such as agricultural residues, waste oils, and microalgae—biodiesel offers a viable alternative to fossil fuels without compromising food supplies or exacerbating the land-use conflicts. This chapter has explored the economic and policy dimensions of biodiesel production, highlighting its potential to decarbonize hard-to-electrify sectors while fostering circular economy principles. Below, we summarize key findings, emphasize the strategic importance of non-food biomass, and outline the actionable recommendations for stakeholders.

### 7.1 Economic and environmental benefits

Biodiesel offers substantial environmental and socioeconomic benefits, making it a key component of sustainable energy strategies. From an ecological perspective, biodiesel derived from waste feedstocks can reduce the lifecycle greenhouse gas emissions by 50–80% compared to conventional petroleum diesel, significantly mitigating climate change impacts. Economically, localized biodiesel production stimulates rural development by generating employment opportunities across the supply chain—from feedstock cultivation and collection to processing and distribution—particularly benefiting agrarian economies. Additionally, biodiesel enhances energy security by diversifying fuel sources and reducing reliance on imported fossil fuels, thereby strengthening national energy independence. Despite these advantages, challenges such as high production costs, feedstock variability, and insufficient infrastructure continue to hinder widespread adoption. Addressing these barriers through technological innovation, policy support, and investment in supply chain development will be crucial to unlocking biodiesel's full potential as a sustainable and economically viable fuel alternative. Biodiesel presents a compelling sustainable energy solution with multiple benefits. Environmentally, it offers the substantial greenhouse gas reductions of 50–80% across its lifecycle compared to petroleum diesel, with waste-derived feedstocks providing the most significant climate benefits. The economic advantages are equally noteworthy, as decentralized biodiesel production stimulates rural economies by creating employment opportunities throughout the value chain—from agricultural feedstock cultivation to processing and distribution networks, particularly benefiting developing regions. From an energy security perspective, biodiesel diversification reduces reliance on volatile global fossil fuel markets while strengthening domestic energy independence. Despite these advantages, key challenges persist, including higher production costs relative to conventional diesel, inconsistent feedstock quality and availability, and underdeveloped distribution infrastructure. Addressing these barriers through technological innovation, policy support, and strategic investments will be critical to realizing biodiesel's full potential as a sustainable transportation fuel alternative. The combination of environmental benefits, rural economic development, and enhanced energy security makes biodiesel an important component of the global transition to cleaner energy systems, though overcoming current limitations remains essential for broader market adoption. Biodiesel production offers significant socioeconomic and strategic benefits, particularly for rural communities and energy security. By establishing localized biodiesel value chains, developing economies can generate employment opportunities across agricultural production, processing facilities, and distribution networks, helping to revitalize rural economies. This decentralized approach to fuel production simultaneously strengthens national energy security by reducing reliance on imported petroleum and insulating countries from volatile global oil markets. The development of domestic biodiesel industries creates a virtuous cycle—farmers gain new markets for oilseed crops, processing plants create skilled manufacturing jobs, and local fuel distribution networks improve energy access in remote areas. However, realizing this potential requires overcoming persistent challenges including high production costs relative to fossil diesel, inconsistent availability of quality feedstocks, and inadequate refining and distribution infrastructure. Strategic investments in agricultural extension programs, processing technology, and fuel logistics will be essential to make biodiesel a viable contributor to both rural development and energy independence goals. When properly supported, biodiesel production can serve as an engine for sustainable economic growth while reducing the energy vulnerabilities faced by many developing nations.

### 7.2 Policy and regulatory frameworks driving biodiesel market growth

Effective policy interventions have proven instrumental in accelerating biodiesel adoption worldwide. Leading biodiesel markets have demonstrated the power of blending mandates, with Indonesia's B35 program (35% biodiesel blend) and Brazil's RenovaBio policy serving as successful models that combine regulatory targets with market mechanisms [17, 50]. Carbon credit systems have emerged as another powerful tool, creating financial incentives that improve biodiesel's economic competitiveness by monetizing its emission reduction benefits. Standardization has played an equally crucial role, with internationally recognized specifications such as ASTM D6751 and EN 14214 ensuring fuel quality, engine compatibility, and consumer confidence. However, the global biodiesel market faces significant challenges due to policy fragmentation. Disparate national regulations, inconsistent subsidy schemes, and varying sustainability criteria create trade barriers and investment uncertainty. This lack of harmonization particularly affects developing countries seeking to establish biodiesel industries, as they struggle to navigate complex export requirements while meeting domestic energy needs. The absence of unified international standards for advanced biofuels and waste-derived feedstocks further complicates market development. Addressing these policy inconsistencies through multilateral cooperation and harmonized sustainability frameworks could unlock significant growth potential in the global biodiesel sector.

## 7.3 Strategic importance of non-food biomass in energy transition

Effective policy interventions have proven instrumental in accelerating biodiesel adoption worldwide. Leading biodiesel markets have demonstrated the power of blending mandates, with Indonesia's B35 program (35% biodiesel blend) and Brazil's RenovaBio policy serving as successful models that combine regulatory targets with market mechanisms. Carbon credit systems have emerged as another powerful tool, creating financial incentives that improve biodiesel's economic competitiveness by monetizing its emission reduction benefits. Standardization has played an equally crucial role, with internationally recognized specifications such as ASTM D6751 and EN 14214 ensuring fuel quality, engine compatibility, and consumer confidence. However, the global biodiesel market faces significant challenges due to policy fragmentation. Disparate national regulations, inconsistent subsidy schemes, and varying sustainability criteria create trade barriers and investment uncertainty. This lack of harmonization particularly affects the developing countries seeking to establish biodiesel industries, as they struggle to navigate complex export requirements while meeting domestic energy needs. The absence of unified international standards for advanced biofuels and waste-derived feedstocks further complicates market development. Addressing these policy inconsistencies through multilateral cooperation and harmonized sustainability frameworks could unlock significant growth potential in the global biodiesel sector.

## 7.4 Call to action for enhanced collaboration among research, industry, and governance

The transition to a sustainable bioeconomy demands unprecedented collaboration among researchers, industries, and policymakers to overcome existing barriers and unlock the full potential of non-food biomass. To drive meaningful progress, stakeholders must establish integrated innovation ecosystems that combine cutting-edge research with scalable industrial applications, supported by forward-looking policies. Researchers should prioritize developing high-yield, climate-resilient feedstocks and efficient conversion technologies, while industries must invest in pilot projects and supply chain infrastructure to bridge the gap between lab-scale innovations and commercial deployment. Governments play a critical role in creating enabling environments through stable regulatory frameworks, financial incentives, and international standards that ensure sustainability and interoperability. A coordinated approach should include knowledge-sharing platforms to disseminate best practices, public-private funding mechanisms to de-risk investments, and workforce development programs to build technical capacity across the value chain. By aligning these efforts through multi-stakeholder partnerships, we can accelerate the commercialization of advanced biofuels, create inclusive economic opportunities in rural communities, and establish biodiesel as a competitive, low-carbon alternative for hard-to-decarbonize sectors. The time for isolated initiatives has passed—only through systemic collaboration can we overcome technical, economic, and regulatory hurdles to make bioenergy a cornerstone of the global energy transition.

## Acknowledgements

The authors gratefully acknowledge the support received from the PYRAGRAF project, financed under the European Union's Horizon Europe framework (Grant Agreement ID 101114608), which provided essential insights into decentralized pyrolytic conversion of agricultural and forestry residues. We also extend our gratitude to the SALTOpower project, financed under the European Union's Horizon Europe framework (Grant Agreement ID 101079303), coordinated by the University of Évora, for their contributions on molten salt-based energy storage and its integration with biomass pyrolysis-driven hydrogen/syngas systems.

## Conflict of interest

The authors declared no conflicts of interest.

## Acronyms and abbreviations

| | |
|---|---|
| ASTM | American Society for Testing and Materials |
| B100 | 100% |
| Biodiesel | pure biodiesel fuel blend stock |
| B20 | 20% biodiesel, 80% |
| Diesel | a common biodiesel blend |
| B30 | 30% biodiesel, 70% |
| Diesel | a biodiesel blend mandate |
| B35 | 35% biodiesel, 65% |
| Diesel | a high-level biodiesel blend mandate |
| CaO | calcium oxide |
| CO | carbon monoxide |
| $CO_2$ | carbon dioxide |
| DES | deep eutectic solvents |
| GHG | greenhouse gas |
| GIS | geographic information system |
| HC | hydrocarbons |
| HVO | hydrotreated vegetable oil |
| IMO | International Maritime Organization |
| ISO | International Organization for Standardization |
| LCA | life cycle assessment |
| NOx | nitrogen oxides |
| OEM | original equipment manufacturer |
| RD&D | research, development, and demonstration |

## Author details

Asma Nouira[1]*, Imen Amri[1], Imene Bekri-Abbes[2] and Diogo Canavarro[1]

1 Renewable Energies Chair, Polo da Mitra da Universidade de Évora,
Nossa Senhora da Tourega, Portugal

2 National Center of Research of Sciences of Materials, Technopole of Borj Cedria,
Tunis, Tunisia

*Address all correspondence to: asma.nouira@uevora.pt; asma.nouira@fst.utm.tn

## References

[1] Jan S, Mishra AK, Bhat MA, Bhat MA, Qyyum MA, Rahman S, et al. Biodiesel in circular economy: A perspective study on recent trends and environmental challenges in sustainable energy production. Renewable Energy. 2025;**255**:123840. DOI: 10.1016/j.renene.2025.123840

[2] Park G, Park J, Kim JY, Lee D, Kwon EE. Use of defatted cottonseed-derived biochar for biodiesel production: A closed-loop approach. Biochar. 2025;7(1):4. DOI: 10.1007/s42773-024-00394-3

[3] Suresh R, Antony JV, Vengalil R, Kochimoolayil GE, Joseph R. Esterification of free fatty acids in non-edible oils using partially sulfonated polystyrene for biodiesel feedstock. Industrial Crops and Products. 2017;**95**:66-74. DOI: 10.1016/j.indcrop.2016.09.060

[4] Burmana AD, Manurung R, Iriany EB, Bentalib AS, Benguerba Y, Affila. Catalyst reuse in biomass-based biodiesel: Life cycle and economic implications for a circular bioenergy economy. Biomass and Bioenergy. 2025;**201**:108071. DOI: 10.1016/j.biombioe.2025.108071

[5] Moreno-Caballero BS, Rivas-García P, Estrada-Baltazar A, Paniagua-Martínez JP, Leveque JM, Cano-Gómez JJ. Exploring the potential for recycling sewage sludge in biodiesel using deep eutectic solvents: A proposal of circular economy based on life cycle assessment approach. Biomass and Bioenergy. 2025;**198**:107845

[6] Devadiga S, Maity S, Mallick N. A circular economy approach for harnessing inorganic and organic carbons coupled with waste remediation using *Chlorella minutissima* for sustainable biodiesel production. Journal of Environmental Chemical Engineering. 2025;**13**(3):117069. DOI: 10.1016/j.jece.2025.117069

[7] Rajak AK, Zeenath SF, Boddula R, Singh S, Varshney D, Sharma K, et al. Efficient application of rice husk ash-derived silica combined with calcined orange peel as catalysts for biodiesel production from Nigella sativa seed oil within a circular economy framework. International Journal of Biological Macromolecules. 2025;**315**(Pt. 1):144067. DOI: 10.1016/j.ijbiomac.2025.144067

[8] Hamza M, Ayoub M, Shamsuddin RB, Mukhtar A, Saqib S, Zahid I, et al. A review on the waste biomass derived catalysts for biodiesel production. Environmental Technology and Innovation. 2021;**21**:101200. DOI: 10.1016/j.eti.2020.101200

[9] Nayak RR, Khairun HS, Singhal R, Bharadwaj AVSL, Gupta NK. Waste-derived catalysts for sustainable biodiesel production: Current status on catalyst development and future prospectives. Journal of the Indian Institute of Science. 2024;**104**(4):841-868

[10] Bhonsle AK, Kumar A, Ray A, Singh J, Rawat N, Atray N. Biodiesel production from used cooking oil at room temperature using novel solvent: A techno-economic perspective, sensitivity analysis and societal implications. Energy Conversion and Management. 2025;**324**:119282. DOI: 10.1016/j.enconman.2024.119282

[11] Varghese SS, Vaisakh PH, Philip E, Paulose TAP, Akhina H. Biofuel as potential candidates for sustainable

development. In: The Food-Energy-Water Triangle for Sustainable Development. 2024. DOI: 10.1201/9781003561453-7

[12] Skarlis S, Kondili E, Kaldellis JK. Small-scale biodiesel production economics: A case study focus on Crete Island. Journal of Cleaner Production. 2012;**20**(1):20-26. DOI: 10.1016/j.jclepro.2011.08.011

[13] Singh SK, Chauhan A, Sarkar B. Strategy planning for sustainable biodiesel supply chain produced from waste animal fat. Sustainable Production and Consumption. 2024;**44**:263-281. DOI: 10.1016/j.spc.2023.10.012

[14] Iddrisu I, Bhattacharyya SC. Ghana's bioenergy policy: Is 20% biofuel integration achievable by 2030? Renewable and Sustainable Energy Reviews. 2015;**43**:32-39. DOI: 10.1016/j.rser.2014.10.066

[15] Sekabira H, Simbeko G, Feleke ST, Manyong VM, Späth L, Krütli P, et al. Biobased product design, development, and utilization in the global south. In: Ogwu MC, Izah SC, Vazquez-Arenas JG, Feleke ST, Wei X, editors. Sustainable Bioeconomy Development in the Global South: Volume II Bioeconomy Techniques. Singapore: Springer; 2025. pp. 95-129. DOI: 10.1007/978-981-96-0305-3_4

[16] Fatima N, Khanam N, Kumari R, Bhattacharya S. Biofuels in India: Policies, reviews, and strategic analysis. BioEnergy Research. 2025;**18**(1):40. DOI: 10.1007/s12155-025-10843-x

[17] Halimatussadiah A, Moeis FR, Siregar AA, Nainggolan D. The potentials of Indonesian biofuel policy's replanting scheme to reduce poverty and enhance regional economy. Land Use Policy. 2025;**157**:107623. DOI: 10.1016/j.landusepol.2025.107623

[18] da Costa MG, Gonçalves DNS, D'Agosto MDA. From commodities to high-value exports: Biofuels as a catalyst for Brazil's energy transition and economic transformation. Sustainable Energy Technologies and Assessments. 2025;**81**:104416. DOI: 10.1016/j.seta.2025.104416

[19] Zardoshti Zadeh Yazdi M, Walther G. Analyzing policy measures for renewable fuel supply chain design in transport sectors. Transportation Research Part D Transport and Environment. 2025;**144**:104789. DOI: 10.1016/j.trd.2025.104789

[20] Schulz V, Bell MGH, Rose JM, Glenn GD. From fossil fuels to biofuels: Circular economy diversification for fossil fuel ports. Transportation Research Part D Transport and Environment. 2025;**144**:104796. DOI: 10.1016/j.trd.2025.104796

[21] Aguilar-Aguilar FA, Mena-Cervantes VY, Hernández-Altamirano R. Analysis of public policies and resources for biodiesel production in México. Biomass and Bioenergy. 2025;**196**:107762. DOI: 10.1016/j.biombioe.2025.107762

[22] Hu Z, Shen J, Tan P, Lou D. Life cycle carbon footprint of biodiesel production from waste cooking oil based on survey data in Shanghai, China. Energy. 2025;**320**:135318. DOI: 10.1016/j.energy.2025.135318

[23] Sulaiman. Present scenario and status of bioenergy sources and production system. In: Singhal RK, Indu ESA, Dwivedi KK, editors. Forage Crops in the Bioenergy Revolution. Singapore: Springer; 2025. DOI: 10.1007/978-981-96-2536-9_3

[24] Silalahi FTR, Simatupang T, Siallagan M. Optimal design for biodiesel supply chain: Case study.

Strategic Planning for Energy and the Environment. 2021;**39**(1):44. DOI: 10.13052/spee1048-4236.39144

[25] Phitthayaphinant P, Nissapa A. Consumers' satisfaction on marketing mix of biodiesel in southern Thailand: A case study in Songkhla, Krabi, and Trang provinces. Kasetsart Journal - Social Sciences. 2011;**32**(1):115-124. Available from: https://so04.tci-thaijo.org/index.php/kjss/article/view/246722

[26] Ezzati F, Babazadeh R, Donyavi A. Optimization of multimodal, multi-period and complex biodiesel supply chain systems: Case study. Renewable Energy Focus. 2018;**26**:81-92. DOI: 10.1016/j.ref.2018.07.005

[27] Díez-Valbuena G, García-Tuero A, Rodríguez E, Hernández Battez A, Díez J. Modeling biodiesel properties by preference learning: Case study of cetane number. Engineering Applications of Artificial Intelligence. 2025;**144**:110107. DOI: 10.1016/j.engappai.2025.110107

[28] Fiorese DA, Gomes LFS, de Souza SNM, Dallmeyer AU, Romano LN. Experimental methodology for assessing the cost of biodiesel production and use biodiesel: A case study of four methyl esters and commercial diesel fuel. Ciencia Rural. 2011;**41**(11):1921-1926. DOI: 10.1590/S0103-84782011001100012

[29] Malani RS, Choudhury HA, Moholkar VS. Waste biorefinery based on waste carbon sources: Case study of biodiesel production using carbon based catalysts and mixed feedstocks of nonedible and waste oils. In: Waste Biorefinery: Integrating Biorefineries for Waste Valorisation. Amsterdam: Elsevier; 2020. pp. 337-378. DOI: 10.1016/B978-0-12-818228-4.00013-7

[30] Purathanang T, Wattana S. An assessment of the implications of biodiesel consumption promotion in the transportation sector for Thailand. IOP Conference Series: Materials Science and Engineering. 2021;**1163**:012015. DOI: 10.1088/1757-899X/1163/1/012015

[31] Geng N, Zhang Y, Sun Y, Geng S. Optimization of biodiesel supply chain produced from waste cooking oil: A case study in China. IOP Conference Series: Earth and Environmental Science. 2019;**264**(1):012006. DOI: 10.1088/1755-1315/264/1/012006

[32] Galas M. Feasibility analysis of a supply chain for castor oil biodiesel distribution in Central Mexico. In: Computer Science and Health Engineering in Health Services, Lecture Notes of the Institute for Computer Sciences, Social Informatics and Telecommunications Engineering. Cham: Springer; 2021. pp. 79-91. DOI: 10.1007/978-3-030-69839-3_6

[33] Zhang F, Wang J, Liu S, Zhang S, Sutherland JW. Integrating GIS with optimization method for a biofuel feedstock supply chain. Biomass and Bioenergy. 2017;**98**:194-205. DOI: 10.1016/j.biombioe.2017.01.004

[34] Thaba SC, Mbohwa C, Pradhan A. Occupational Health and Safety in the Biofuels Industry: The Case Study of Small Scale Biodiesel Plants in Gauteng Province, South Africa. 2015. Available from: https://hdl.handle.net/10210/86191

[35] Watanabe K, Bijman J, Slingerland M. Institutional arrangements in the emerging biodiesel industry: Case studies from Minas Gerais—Brazil. Energy Policy. 2012;**40**:381-389. DOI: 10.1016/j.enpol.2011.10.023

[36] de Souza J, Carmo Andrade L, de Rezende D. Case study: Production

of biodiesel from frying oil for use in employees transport minibuses inside Fiat's Plant in Brazil. In: SAE Technical Paper 2014-36-0156. 2014. DOI: 10.4271/2014-36-0156

[37] Carrero A, Pérez Á. Advances in biodiesel quality control, characterisation and standards development. In: Luque R, Melero JA, editors. Advances in Biodiesel Production: Processes and Technologies. Cambridge: Woodhead Publishing; 2012. pp. 91-112. DOI: 10.1533/9780857095862.1.91

[38] Clampitt BH, Sardisco JB. Calibration standards for biodiesel fuels. Journal of ASTM Internationa. 2010;7(3):1-7. DOI: 10.1520/JAI102430

[39] Fernando S, Hanna M. Issues with biodiesel purity and the current US biodiesel standard. In: AIChE Annual Meeting Conference Proceedings; 2005 Nov 13-18; Cincinnati, OH. Vol. 2005. New York: American Institute of Chemical Engineers. p. 5412

[40] Hoekman SK, Broch A, Robbins C, Ceniceros E, Natarajan M. Review of biodiesel composition, properties, and specifications. Renewable and Sustainable Energy Reviews. 2012;**16**(1):143-169. DOI: 10.1016/j.rser.2011.07.143

[41] More G, Rao YVH. Biodiesel production with the help of different additives on the basis of standards – A review. Journal of Advanced Research in Dynamical and Control Systems. 2018;**10**(9):2050-2064

[42] Rajamohan S, Kasimani R, Ramakrishnan P, Peer MS. A review on the properties, performance and emission aspects of the third generation biodiesels. Renewable and Sustainable Energy Reviews. Nov 2017;**82**(Pt 3):2970-2992. DOI: 10.1016/j.rser.2017.10.037

[43] Howell S. U.S. biodiesel standards - An update of current activities. In: SAE Tech Pap. 971687. 1997. DOI: 10.4271/971687

[44] Fortin E. Repoliticising multi-stakeholder standards processes: The roundtable on sustainable biomaterials' standards and certification scheme. Journal of Peasant Studies. 2017;**45**(5):1-20. DOI: 10.1080/03066150.2017.1285286

[45] Pasimura I, Afrizal A, Novarino W. Impact of Indonesian sustainable palm oil (ISPO) certification to environmental behavior of palm oil plantation companies. International Journal of Agriculture System. 2022;**10**(1):26. DOI: 10.20956/ijas.v10i1.3254

[46] Xu YJ, Li GX, Sun ZY. Development of biodiesel industry in China: Upon the terms of production and consumption. Renewable and Sustainable Energy Reviews. 2015;**54**:318-330. DOI: 10.1016/j.rser.2015.10.035

[47] Febriansyah H, Yayienda NF, Mulyono AB, Isharyadi F. Water contents and monoglycerides as development role of biodiesel standard in Indonesia for B30 implementation. In: Proceedings of the International Conference on Electrical Engineering, Computer Science and Informatics (EECSI); 2020 Sep 27-28; Malang, Indonesia. Yogyakarta: EECSI; 2020. pp. 50-55. DOI: 10.23919/EECSI50178.2020.9232480

[48] Braun JV, Santos VOB, Fontoura L, et al. GC-FID methodology validation for the fatty esters content determination in biodiesel with hexadecyl acetate as the internal standard. Quimica Nova. 2017;**40**(9):1043-1050. DOI: 10.21577/0100-4042.20170103

[49] Rodriguez Coronado C, Tuna CE, Zanzi R, Vane LF, Silveira JL. Development of a thermoeconomic methodology for optimizing biodiesel

production. Part II: Manufacture exergetic cost and biodiesel production cost incorporating carbon credits, A Brazilian case study. Renewable and Sustainable Energy Reviews. 2014;**29**:565-572. DOI: 10.1016/j.rser.2013.08.064

[50] Tiburcio RS, Macêdo TR, Neto AMP. Brazilian biofuels policy (RenovaBio): Overview and generation of decarbonization credits by biodiesel production facilities. Energy for Sustainable Development. 2023;**77**:101334. DOI: 10.1016/j.esd.2023.101334

[51] Soares FR, Martins G, Seo ESM. An assessment of the economic aspects of $CO_2$ sequestration in a route for biodiesel production from microalgae. Environmental Technology. 2013;**34**(13-16):1777-1781. DOI: 10.1080/09593330.2013.816784

[52] Branco JEH, Prates RC, Oliveira ALDN, Carvalho AKFD. Evaluation of the greenhouse gas emissions of the Brazilian biodiesel and the impact of the mandatory blending of the biofuel into commercial diesel fuel. Chemosphere. 2025;**376**:144268. DOI: 10.1016/j.chemosphere.2025.144268

[53] Zheng F, Cho HM. Study on biodiesel production: Feedstock evolution, catalyst selection, and influencing factors analysis. Energies. 2025;**18**(10):2533. DOI: 10.3390/en18102533

[54] Custom Market Insights. Europe biofuel market [Internet]. 2024. Report No.: CMI71418. Available from: https://www.custommarketinsights.com/report/europe-biofuel-market/ [Accessed: November 25, 2025]

[55] Bockey D. The significance and perspective of biodiesel production – A European and global view. Oilseeds & Fats Crops and Lipids. 2019;**26**:40. DOI: 10.1051/ocl/2019042

[56] Simionescu M, Bilan Y, Streimikiene D. The impact of biodiesel consumption by transport on economic growth in the European Union. Inzinerine Ekonomika-Engineering Economics. 2019;**30**(1):50-58. DOI: 10.5755/j01.ee.30.1.21831

[57] Lundberg L, Cintas Sanchez O, Zetterholm J. The impact of blending mandates on biofuel consumption, production, emission reductions and fuel prices. Energy Policy. 2023;**183**:113835. DOI: 10.1016/j.enpol.2023.113835

[58] Gasol CM, Salvia J, Serra J, Antón A, Sevigne E, Rieradevall J, et al. A life cycle assessment of biodiesel production from winter rape grown in southern Europe. Biomass and Bioenergy. 2012;**40**:71-81. DOI: 10.1016/j.biombioe.2012.02.003

[59] Omoniyi A, Vadehra M, Thompson C, Braslavets S, Duggan S, Hensley A. A comprehensive analysis of the process, economics, and business model of converting used cooking oil to biodiesel in the Northeastern United States. Journal of Renewable and Sustainable Energy. 2025;**17**(3):033101. DOI: 10.1063/5.0259200

[60] Whistance J, Ripplinger D, Thompson W. Biofuel-related price transmission using renewable identification number prices to signal mandate regime. Energy Economics. 2016;**55**:19-29. DOI: 10.1016/j.eneco.2015.12.026

[61] Yadala S, Smith JD, Young D, Crunkleton DW, Cremaschi S. Optimization of the algal biomass to biodiesel supply chain: Case studies of the state of Oklahoma and the United

States. Processes (Basel). 2020;**8**(4):476. DOI: 10.3390/pr8040476

[62] Debnath D, Whistance J, Thompson W. The U.S. biodiesel outlook: A pathway toward sustainability. Sustainability (Basel). 2025;**17**(2):689. DOI: 10.3390/su17020689

[63] Tanaka T, Guo J, Wang X. Price interconnection of fuel and food markets: Evidence from biodiesel in the United States. GCB Bioenergy. 2023;**15**(7):886-899. DOI: 10.1111/gcbb.13052

[64] National Biodiesel Board. Fuel quality policy. In: Dahlquist E, editor. Biomass to Biofuels. London: Academic Press; 2015. pp. 541-542. DOI: 10.1016/B978-0-12-407909-0.00032-8

[65] McCormick R, Alleman T, Nelson R. Statistical treatise on critical biodiesel (B100) quality properties in the United States from 2004-2022. In: SAE Technical Paper 2023-01-1218. Warrendale, PA: SAE International; 2023. DOI: 10.4271/2023-01-1218

[66] Moser BR. Biodiesel production, properties, and feedstocks. In Vitro Cellular & Developmental Biology. Plant. 2009;**45**(3):229-266. DOI: 10.1007/s11627-009-9204-z

[67] Go AW, Agapay RC, Ju Y-H, Angkawijaya AE, Tran Nguyen PL. The capacity of major countries in Southeast Asia in meeting biodiesel mandates and pursue higher blends based on available major feedstocks. Biofuels. 2022;**13**(5):601-613. DOI: 10.1080/17597269.2020.1824474

[68] Mukherjee I, Sovacool BK. Palm oil-based biofuels and sustainability in Southeast Asia: A review of Indonesia, Malaysia, and Thailand. Renewable and Sustainable Energy Reviews. 2014;**37**:1-12. DOI: 10.1016/j.rser.2014.05.001

[69] Papilo P, Marimin M, Hambali E, Machfud M, Yani M, Asrol M, et al. Palm oil-based bioenergy sustainability and policy in Indonesia and Malaysia: A systematic review and future agendas. Heliyon. 2022;**8**(10):e10919. DOI: 10.1016/j.heliyon.2022.e10919

[70] Shi X, Goto S. Harmonizing biodiesel fuel standards in East Asia: Motivations, progress and challenges. Applied Energy. 2013;**105**:217-222. DOI: 10.1016/j.apenergy.2013.01.001

[71] Zhu Y, Xu J, Li Q, Mortimer PE. Investigation of rubber seed yield in Xishuangbanna and estimation of rubber seed oil based biodiesel potential in Southeast Asia. Energy. 2014;**69**:837-842. DOI: 10.1016/j.energy.2014.03.085

[72] Krukowski A, Kijek T, Nowak A, et al. The potential of the European Union countries to produce biomass for biodiesel production and consumption purposes. In: Proceedings of the 8th International Scientific Conference Rural Development 2017; 2017 Nov 23-24; Kaunas, Lithuania. Kaunas: Aleksandras Stulginskis University; 2017. pp. 839-844. DOI: 10.15544/RD.2017.169

[73] Mercado-Cordova J. Biofuel technologies and global decarbonization by 2050: The role of technological learning, spillovers effects and feedstock constraints. Energy Strategy Reviews. 2025;**62**:101932. DOI: 10.1016/j.esr.2025.101932

[74] Aryanfar Y, Keçebaş A, Sadabad AN, Fernandez JB, Wu W. Production of biodiesel from industrial sludge: Recent progress, challenges, perspective. PRO. 2024;**12**(11):2517

[75] Panda S, Biswas CK, Paul S. Production of biodiesel from industrial sludge: Recent Progress, challenges,

perspective. In: Recent Trends in Management and Utilization of Industrial Sludge. 2024. pp. 337-357

[76] Abomohra AE-F, Elsayed M, Esakkimuthu S, El-Sheekh M, Hanelt D. Potential of fat, oil and grease (FOG) for biodiesel production: A critical review on the recent progress and future perspectives. Progress in Energy and Combustion Science. 2020;**81**:100868

[77] Stamenković OS, Gautam K, Singla-Pareek SL, Pareek A, Veljković VB. Biodiesel production from camelina oil: Present status and future perspectives. Food and Energy Security. 2023;**12**(1):e340

[78] D'ambrosio V, Scelsi E, Pastore C. Non-edible oils for biodiesel production: State of the art and future perspectives. In: Biodiesel Production Feedstocks Catalysts and Technologies. 2022. pp. 49-66

[79] Ma Y, Gao Z, Wang Q, Liu Y. Biodiesels from microbial oils: Opportunity and challenges. Bioresource Technology. 2018;**263**:631-641

[80] Mondal AS, Sharma R, Trivedi N. Cutting-edge approaches for overcoming challenges in microalgal biodiesel production. In: Microalgal Biomass for Bioenergy Applications. 2023. pp. 355-394

[81] Singh CS, Kumar N, Gautam R. Supercritical transesterification route for biodiesel production: Effect of parameters on yield and future perspectives. Environmental Progress & Sustainable Energy. 2021;**40**(6):e13685

[82] Marinković DM, Stanković MV, Veličković AV, Veljković VB, Jovanović DM. Calcium oxide as a promising heterogeneous catalyst for biodiesel production: Current state and perspectives. Renewable and Sustainable Energy Reviews. 2016;**56**:1387-1408

[83] Fauzi AHM, Amin NAS. Catalysis in biodiesel synthesis: Challenges and future perspectives. In: Advances in Biofuels. 2013. pp. 127-152

[84] Ismaeel HK, Albayati TM, Dhahad HA, Saady NMC, Zendehboudi S. Strategies for biodiesel production with the role of reactor technologies: A comprehensive review. Chemical Engineering and Processing - Process Intensification. 2024;**200**:109767

[85] Julio AAV, Milessi TS, Ocampo Batlle EA, Yepes Maya DM, Escobar Palacio JC. Techno-economic and environmental potential of renewable diesel as complementation for diesel and biodiesel in Brazil: A comprehensive review and perspectives. Journal of Cleaner Production. 2022;**371**:133431

[86] Ma Y, Liu Y. Biodiesel production: Status and perspectives. In: Biomass Biofuels Biochemicals Biofuels Alternative Feedstocks and Conversion Processes for the Production of Liquid and Gaseous Biofuels. 2019. pp. 503-522

[87] Kokkinos NC, Emmanouilidou E, Sharma SK. Waste-to-biofuel production for the transportation sector. In: Energy Environment and Sustainability. 2024. pp. 99-122

[88] Foteinis S, Chatzisymeon E, Litinas A, Tsoutsos T. Used-cooking-oil biodiesel: Life cycle assessment and comparison with first- and third-generation biofuel. Renewable Energy. 2020;**153**:588-600

[89] Vasileiadis V, Zafeiriou E, Karmalis I, Karayannis V, Tsanaktsidis C. Valorization of biodiesel from waste vegetable and animal resources: Advancing circular bioenergy through sustainable automotive and marine fuel

blends. International Journal of Energy Sector Management. 2025

[90] Sevim C, Zincir B. Biodiesel and renewable diesel as a drop-in fuel for decarbonized maritime transportation. In: Agarwal AK, Valera H, editors. Potential and Challenges of Low Carbon Fuels for Sustainable Transport. Energy, Environment, and Sustainability. Singapore: Springer; 2022. DOI: 10.1007/978-981-16-8414-2_10

[91] Carlson NA, Talmadge MS, Tan ECD, Newes EK, McCormick RL. Refinery perspective on decarbonizing with marine biofuels. Energy & Fuels. 2023;**37**(19):14411-14420

[92] Sonthalia A, Kumar N. Performance improvement and emission reduction potential of blends of hydrotreated used cooking oil, biodiesel and diesel in a compression ignition engine. Energies. 2023;**16**(21):7431

[93] Zandie M, Ng HK, Gan S, Said MFM, Cheng X. The viability of using gasoline-integrated biodiesel-diesel mixtures in engines as a solution to greenhouse gas emissions: A review. Clean Energy. 2022;**6**(6):848-868

[94] Suresh R, Wang Y, Xu C, Fu X, Aggarwal SK. A numerical study on combustion and emissions characteristics of diesel/biodiesel blends in a locomotive engine. The American Society of Mechanical Engineers Internal Combustion Engine Division Fall Technical Conference Proceedings Series. 2024;**2024**:V001T06A011

[95] León JA, Montero G, Coronado MA, Quintana L, Armenta JM. Thermodynamic analysis of waste vegetable oil conversion to biodiesel with solar energy. Energies. 2022;**15**(5):1834

[96] Guangqian D, Bekhrad K, Azarikhah P, Maleki A. A hybrid algorithm based optimization on modeling of grid independent biodiesel-based hybrid solar/wind systems. Renewable Energy. 2018;**122**:551-560

Section 3

# Advanced Technological Treatment Pathways

Chapter 6

# Environmental Sustainability through the Bioremediation of Petroleum Pollutants Using Gamma Radiation

*Razia Tahseen, Muhammad Ravish Khan and Muhammad Imran*

## Abstract

Petroleum contamination harms the environment and human health because of its persistence, toxicity, and carcinogenic nature. Traditional remediation methods, including physicochemical treatments, are often incompetent, costly, or result in additional pollution. Bioremediation through microbial degradation presents a more sustainable option but often lacks efficiency due to low metabolic rates and set environmental conditions. This chapter focuses on the new idea of using gamma radiation to enhance bacterial activity for petroleum bioremediation. Gamma irradiation makes microbial enzymatic activity, stress resistance, and hydrocarbon degradation pathways more favorable through genetic mutations. Additionally, the hydrocarbons are made more accessible for microbial uptake through bio-oxidation and fragmentation, increasing the likelihood of altered hydrocarbon structures. Irradiated bacterial strains, particularly mutant 300-B, are recognized to demonstrate greater efficiency in degradation processes than their wild-type counterparts. Soil microcosm studies demonstrate that the gamma-irradiated bacteria increased the rate of petroleum hydrocarbons degradation and, under optimal conditions, can yield up to 53.45% total petroleum hydrocarbon (TPH) removal as compared to the wild strain (30%). Increased hydrocarbon mineralization possibly through upregulated enzymes such as alkane hydroxylases and dioxygenases. This chapter discusses the primary components of the processes of bacteria enhancement by gamma radiation, optimal dosage radiation strategies, and the molecular mechanisms of gamma-assisted bioremediation. Although there are issues such as optimization of dose and the possible lethal effects on beneficial microbes from high-radiation levels as well as other environmental safety issues, gamma-assisted bioremediation is an attractive and easily implemented strategy to remove petroleum contamination.

**Keywords:** bacteria, gamma radiation, bioremediation, petroleum contaminants, environmental sustainability

## 1. Introduction

Petroleum is made up of multiple complex hydrocarbon-derived compounds from fossil fuel products. Chemically speaking, it is a complex mixture of organic compounds mainly composed of hydrogen and carbon atoms, with trace amounts of nitrogen, oxygen, sulfur, and metallic elements. The four fractions of petroleum-derived hydrocarbons are the asphaltene fraction (phenols, fatty acids, ketones, esters, and porphyrins), the aliphatic fraction (saturates), the aromatic fraction, and the resins (pyridines, quinolines, carbazoles, sulfoxides, and amides) [1]. Each of these groups is described below in **Figure 1**.

The primary constituents of crude oil and petroleum products are aliphatic hydrocarbons. Based on their chemical structure, saturates can be categorized into cycloalkanes (naphthene) and alkanes (paraffin). Aliphatic hydrocarbons differ from aromatic hydrocarbons, methane derivatives, and non-aromatic and non-cyclic organic molecules in saturated and unsaturated linear or branched open-chain structures. Two to six aromatic rings (such as a fused benzene ring) with or without alkyl substituents arranged in linear, cluster, or angular configurations within their structure are how aromatic fractions of organic compounds are categorized. The characteristics of resins (pyridines, quinolines, carbazoles, sulfoxides, and amides) and asphaltenes (phenols, fatty acids, ketones, esters, and porphyrins) affect the behavior during production and refinement [2].

The environment becomes contaminated through spills, industrial discharges, and improper disposal methods, seriously affecting ecosystems and human health [3]. The endurance of organic pollutants, which tend to accumulate in the biological food chain, hurts biodiversity and lowers environmental quality because of their highly

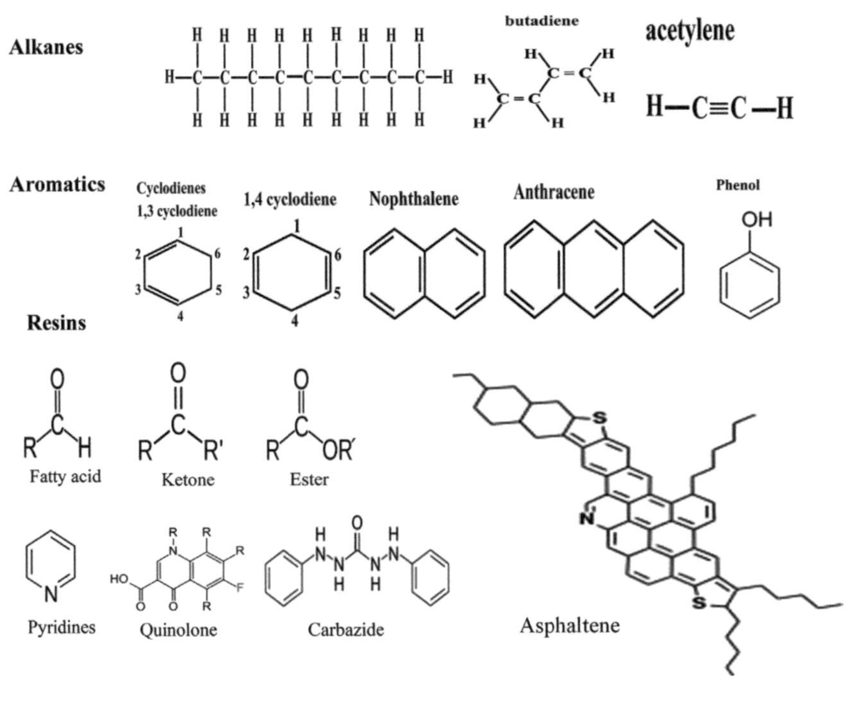


**Figure 1.**
*Constituents of crude oil categorized into four classes hydrocarbons.*

toxic, carcinogenic, mutagenic, and teratogenic properties. The lethal effects of the petroleum hydrocarbons are shown in **Figure 2**.

Petroleum contamination in the soil also negatively affects seed germination, decreasing photosynthetic pigments while slowing absorption, inhibiting root growth, and causing plant leaf defects and cell damage. The harmful effects of petroleum on plants include biological membrane destruction, impairment of signaling in metabolic pathways, and deformation of root structures [4].

Multiple studies prove that plant cells accept low molecular weight hydrocarbons, which destroy the complete plant structure. A range of physicochemical techniques, such as coagulation, flocculation, UV degradation, adsorption, ion exchange, electrochemical techniques, membrane filtration, ozonation, and photocatalysis, were applied to remove traces of organic pollutants from the biosphere. All of these techniques have some shortcomings, including failure to remove organic constituents, excessive space requirements for biological treatment, secondary pollution resulting from the spatial or modal transfer of contaminants after adsorption or precipitation, high-energy costs, steep expenditure, and low longevity of membranes or membrane filtration systems.

Researchers define bioremediation as "biodegradation through the biological process, simplifying chemical compounds." Bioremediation is a promising and eco-friendly approach to mitigating petroleum pollution in contaminated environments. This method employs microorganisms, such as bacteria, to degrade hydrocarbons into less harmful compounds. The microbial degradation of the organic compounds produces minimally complex end-products that resolve their chemical defects [5].

**Figure 2.**
*Effect of organic pollutants on living organisms.*

The traditional approach involving microbial decomposition encounters two main drawbacks: it happens at a slow rate and shows poor effectiveness against challenging pollutants. The efficiency of natural microbial populations can be limited by environmental factors and the metabolic capabilities of microbes [6]. Therefore, the breakdown of organic petroleum and different aromatic compounds in the environment requires sophisticated methods.

Gamma-ray irradiation has emerged as a potential tool to enhance bacterial performance by inducing beneficial mutations, increasing tolerance to pollutants, and improving metabolic pathways for hydrocarbon degradation. This chapter explores the application of gamma irradiation to bacteria, its effects on microbial physiology, its impact on the bioremediation of petroleum pollutants, and ultimately enhancing environmental sustainability. Additionally, gamma radiation has been found to induce dehydrogenation and carboxylation reactions of hydrocarbons, which further enhance microbial accessibility. The formation of new oxygenated functional groups enables enzymatic attack by oxidative microbes, thereby expediting biodegradation (**Figure 3**).

Ionizing radiation, called gamma radiation, appears to be a modern technology for environmental safety, as shown in **Figure 4**. Gamma radiation functions with

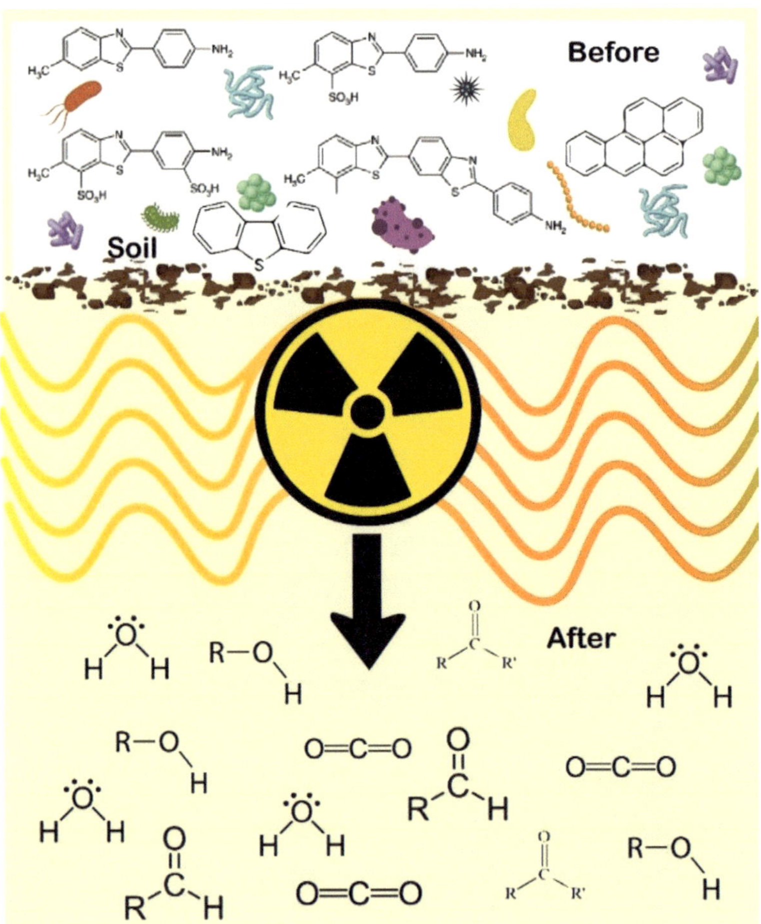


**Figure 3.**
*Use of gamma radiation in environmental sustainability.*

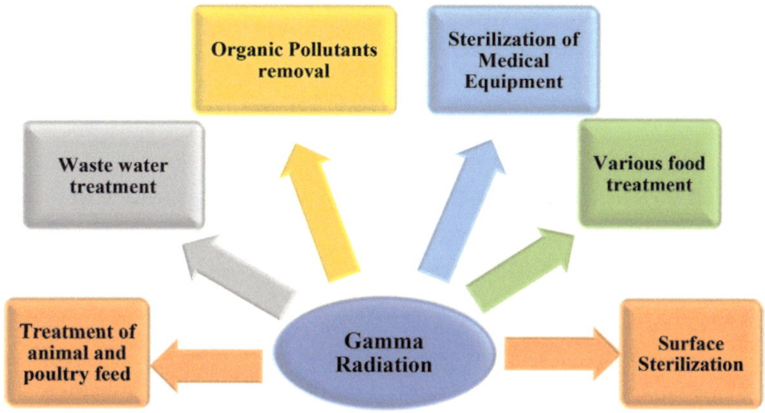


**Figure 4.**
*Use of gamma radiations for environmental safety.*

biological processes by delivering combined benefits to microorganisms through mutation-based enhancement and hydrocarbon substance degradation. For example, irradiated *Alcanivorax* spp. in marine spills to avoid risks associated with invasive species.

## 2. Fundamentals of gamma-ray irradiation

Gamma rays are emitted from radioactive isotopes like cobalt-60 (Co-60) and cesium-137 (Cs-137). These isotopes and other emitters of high-energy electromagnetic waves suffer mutations to their genetic structure because they can penetrate biological tissues [7]. In bacterial cells, exposure to gamma radiation can:

i. Induce mutations that enhance metabolic pathways for hydrocarbon degradation.

ii. Increase bacterial resistance to toxic petroleum components.

iii. Improve survival rates in harsh environmental conditions.

iv. Enhance enzymatic functions, leading to faster breakdown of hydrocarbons.

v. Stimulate adaptive responses, allowing microbes to thrive in contaminated sites.

## 3. Mechanism of gamma-ray-induced mutations in microbes

i. Gamma rays interact with cellular components, leading to DNA damage and repair processes that introduce mutations. These mutations can result in enhanced enzymatic activity, improved stress resistance, and increased adaptability to hydrocarbon-rich environments. These processes reduce molecular weight and increase functionalized hydrocarbons, facilitating microbial uptake and metabolism.

ii. Additionally, gamma radiation has been found to induce dehydrogenation and carboxylation reactions, which further enhance microbial accessibility. The formation of new oxygenated functional groups enables enzymatic attack by oxidative microbes, thereby expediting biodegradation.

## 4. Exposure of gamma radiation to hydrocarbon degradation

Gamma radiation interacts with hydrocarbons through direct and indirect mechanisms:

i. *Direct interaction:* High-energy photons break carbon-carbon and carbon-hydrogen bonds, leading to fragmentation of hydrocarbon molecules. The intense electromagnetic energy of gamma rays performs radiolysis by converting complex organic materials into less toxic and essential products such as $CO_2$ and $H_2O$ [8].

ii. *Indirect interaction:* Ionization of water molecules generates reactive oxygen species (ROS), such as hydroxyl radicals and hydrogen peroxide, which oxidize hydrocarbons, making them more accessible to microbial degradation, as shown in **Figure 5** [9].

## 5. Dose optimization for bioremediation

Low to moderate doses (0.1–5 kGy) induce beneficial mutations without significantly impairing the viability of microorganisms [10]. Higher doses (>10 kGy) can be detrimental, leading to cell death and loss of functional traits. Radiation doses must be carefully optimized to balance genetic improvement with cell viability, typically between 100 and 1000 Gy. High-dose can be lethal; exposure can lead to excessive DNA damage, reducing bacterial efficiency, and controlled irradiation can enhance microbial traits beneficial for bioremediation [11].

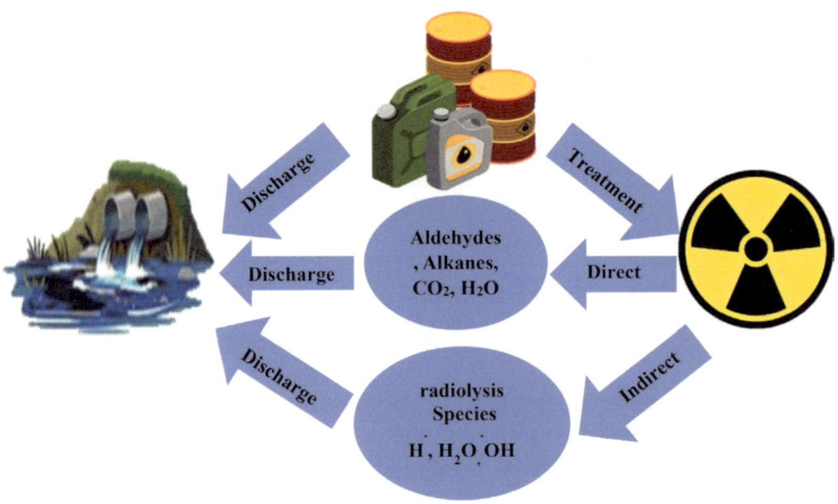


**Figure 5.**
*Mechanism of gamma radiation on hydrocarbon degradation.*

## 6. Microbial responses to gamma radiation

Microbial consortia, including bacteria and fungi, play a pivotal role in petroleum biodegradation. Gamma radiation alters hydrocarbon properties, leading to:

i. *Enhanced microbial enzymatic activity:*

ii. Oxidized and fragmented hydrocarbons are more susceptible to enzymatic breakdown.

iii. Increased microbial growth:

iv. Gamma irradiation promotes microbial proliferation by reducing toxicity and increasing bioavailability.

v. Induction of stress-response mechanisms:

vi. Some microbes exhibit increased resistance and adaptability to radiation-induced hydrocarbon modifications.

Key microbial genera such as *Pseudomonas*, *Bacillus*, *Acinetobacter*, and *Mycobacterium* have shown enhanced degradation of irradiated petroleum hydrocarbons. Studies indicate that when exposed to irradiated hydrocarbons, these microbes produce higher concentrations of hydrocarbon-degrading enzymes such as monooxygenases and dioxygenases [11, 12]. This enhanced enzymatic activity leads to faster mineralization and detoxification of petroleum contaminants. The process of gamma stress results in the selection of strains that feature stronger hydrocarbon degradation genes such as *alkB* and *nahAc*. The irradiated *P. putida* populations degrade 30% more crude oil than wild strains within 14 days [13].

## 7. Bioremediation of petroleum pollutants

Bioremediation leverages microorganisms, plants, or enzymes to detoxify petroleum-contaminated environments. Bioaugmentation introduces hydrocarbon-degrading microbes (e.g., *Pseudomonas*, *Acinetobacter*) to polluted sites. Biostimulation enhances microbial activity via nutrient amendments (e.g., nitrogen and phosphorus) [14]. Due to their structural complexity, petroleum hydrocarbons vary in biodegradability: alkanes degrade faster than aromatic or asphaltene compounds. For instance, n-alkanes ($C_{10}$–$C_{25}$) are metabolized within weeks, while polycyclic aromatic hydrocarbons (PAHs) like benzo[a] pyrene may persist for decades. Challenges include oxygen limitation in anaerobic environments, nutrient scarcity, and toxicity of hydrocarbon byproducts. Gamma radiation addresses these issues by pre-conditioning pollutants and microbes, improving bioavailability and enzymatic efficiency.

### 7.1 Optimal radiation doses vary by pollutant and microbial strain

Low doses (1–3 kGy) enhance enzyme activity, while higher doses (>10 kGy) risk microbial mortality. A meta-analysis identified 2–5 kGy as ideal for mixed microbial communities, maximizing mutation rates without significant biomass loss. Temperature (20–35°C), pH (6–8), and nutrient availability (C: N: P ratio = 100:10:1) critically influence outcomes. Bioaugmentation with irradiated strains and biostimulation using nitrogen fertilizers synergizes with gamma treatment, achieving 90% TPH (total petroleum hydrocarbons) removal in marine sediments [15].

A trial in the Gulf of Mexico used gamma-irradiated (3 kGy) *Alcanivorax borkumensis* to degrade 70% of stranded oil in 4 months, outperforming traditional dispersants [16]. At a former refinery site in India, five key pretreatment and biostimulation methods reduced TPH from 12,000 to 1200 mg/kg in 8 months, meeting regulatory standards [17]. Soil heterogeneity required site-specific radiation dosing, increasing costs by 20%. Gamma-assisted bioremediation in Alaska achieved 55% diesel degradation at 5°C, compared to <10% in unirradiated plots by cold-adapted *Psychrobacter* strains.

Recent lab-scale experiments demonstrate gamma radiation's potential to optimize bioremediation. *Rhodococcus erythropolis* exposed to 2.5 Gy gamma rays showed a 50% increase in *alkB* gene expression, enhancing hexadecane degradation by 65% in 10 days [18]. Gamma irradiation (5 Gy) of crude oil reduced its viscosity by 40%, enabling *Aspergillus flavus* to degrade 80% of PAHs in 21 days, compared to 45% in untreated controls [19].

Combining gamma radiation (3 kGy) with biochar in diesel-contaminated soil increased microbial diversity and accelerated TPH removal by 2.5-fold. These studies highlight dose-dependent effects; low doses (1–5 kGy) enhance microbial activity, while higher doses (>10 kGy) may inhibit growth due to DNA damage.

### 7.2 Gamma radiation significantly accelerates degradation kinetics

Gamma radiation increases the degradation speed; in irradiated marine sediments, n-C18 degradation rates increased from 0.12 to 0.35 mg/kg/day. Pretreatment with 4 kGy reduced pyrene's half-life from 120 to 28 days in soil, attributed to hydroxyl radical-mediated cleavage. Asphaltenes, typically resistant to biodegradation, showed 30% degradation after 60 days when irradiated (7 Gy) and inoculated with *Bacillus subtilis* [20]. Mixed communities outperform monocultures; irradiated *Pseudomonas-Streptomyces* consortia degraded 95% of crude oil in 30 days. Aerobic systems achieve faster degradation than anaerobic setups due to oxygen-dependent oxygenase [21].

### 7.3 Molecular insights into gamma-assisted bioremediation

Gamma radiation induces molecular changes in microbial communities and hydrocarbon structures, enabling efficient bioremediation.

## 8. Microbial genetic adaptations

### 8.1 DNA repair mechanisms

Radiation-resistant microbes like *Deinococcus radiodurans* employ *RecA*-mediated homologous recombination and nucleotide excision repair to fix double-strand breaks. These pathways upregulate stress-response genes (e.g., *katA* for catalase) that mitigate oxidative damage. Gamma photons induce single- and double-strand DNA breaks, triggering error-prone repair systems like (save our souls) SOS response.

This mutagenesis can upregulate catabolic enzymes, such as oxygenase and dehydrogenases, which oxidize hydrocarbons into water-soluble intermediates. For instance, laccase activity in *Aspergillus niger* increases by 40% after two key exposures, enhancing PAH breakdown [22].

### 8.2 Catabolic gene induction

Sublethal gamma doses (1–5 kGy) activate promoters for hydrocarbon degradation genes. For example, *alkB* (alkane hydroxylase) in *Pseudomonas* spp. is overexpressed by 200% under gamma stress, enhancing hexadecane metabolism [23]. Gamma photons (energy ~1.25 MeV) generate hydroxyl radicals ($^{\bullet}OH$) via water radiolysis, which oxidize PAHs into quinones and carboxylic acids [24]. For instance, 5 kGy irradiation cleaves benzo[a] pyrene's aromatic rings, reducing its carcinogenicity by 80% [25]. Heavy hydrocarbons like asphaltenes undergo desulfurization and dementalization, breaking metal-porphyrin complexes that inhibit microbial uptake. Irradiated microbes activate SoxRS and OxyR systems to neutralize reactive oxygen species (ROS), preventing cellular damage while promoting hydrocarbon oxidation.

### 8.3 Extremophile enzymes

Gamma irradiation selects for extremophiles producing robust enzymes like laccases and cytochrome P450 oxygenase, which remain active in high-radiation, high-salinity environments. Under aerobic conditions, most organic pollutants undergo subsequent rapid and complete degradation [26]. Oxidation and activation are the first intracellular attacks that occur because of oxidation and enzymatic motivation by peroxides and oxygenates, and oxygen implementation is crucial.

Organic pollutants are dissolved into soluble, peripheral intermediary stages that can be used for intermediate catabolism, such as the tricarboxylic acid cycle [27]. Metabolites of major precursors, acetyl-CoA, pyruvate, and succinate, produce the cell biomass metabolism. Gluconeogenesis synthesizes the saccharides that are required for various biosynthesis and growth. This could be brought about through a particular enzyme system for pH reduction. Other processes, such as microbial cell adhesion to substrates and the formation of biosurfactants, are also involved. The consortium has proven that there are different mechanisms to metabolize or degrade petroleum hydrocarbons than simply what individual cultures can use. Pilot studies reveal heterogeneous soil matrices and radiation safety as key hurdles [28].

## 9. Bioremediation of soil contaminated with petroleum using gamma-irradiated bacteria—A case study

### 9.1 Irradiation of bacteria strain 33 by gamma rays

Shake flasks batch culture in nutrient broth were centrifuged at 8000 rpm (10,000 G) for 15 minutes. The cells were washed and suspended in normal saline to an OD of 0.7 at 600 nm wavelength and transferred to 30 ml McCartney vials. The cells were exposed to 200, 300, 400, and 500 Gy gamma radiation in a $^{60}$Co irradiator (Mark-IV, USA) available at the Nuclear Institute of Agriculture and Biology (NIAB),

Faisalabad. At the exposure rate of 40 Gy/hour. These treated cultures were irradiated to three log kills [29].

### 9.2 Test for three log kill

To find the best dose (with three log kill) for mutation, the exposed cells to gamma-ray doses, that is, 200, 300, 400, and 500 Cy, were serially diluted, plated on nutrient agar plates, and incubated at 37°C. After counting the colonies, the survival rate was checked, and three log kills were observed at the dose level of 300 Gy, and the strain was named 300-B as shown in **Figure 6**.

The graph illustrates a negative linear trend, with bacterial survival (measured in Log CFU/ml) decreasing as the radiation dosage (Gray) increases from 0 to 500 Gy. Larger radiation doses significantly impair bacterial viability, with an exponential decay rate. The effect of gamma radiation on bacterial survivability demonstrates that as radiation dosage increases, bacterial viability drops dramatically. This indicates a robust relationship between radiation exposure and bacterial cell death, with a logarithmic decline trend.

### 9.3 The bioremediation of petroleum-contaminated soil

This research aimed to determine how effective the mutant strain (300-B) is compared to its parent strain (33) in bioremediating petroleum-contaminated soil in the local environment. Several environmental factors, such as soil moisture content, nitrogen, phosphorus, and potassium metabolism relative to the consumption of hydrocarbons within the soil microcosms, were studied. The assessment of the parent and mutant strains centered on estimating their biodegradation potential concerning the concentration of petroleum hydrocarbons in the soil.

The soil's physicochemical and biological properties were analyzed before and after biodegradation. The petrophysical and ecological composition of the native

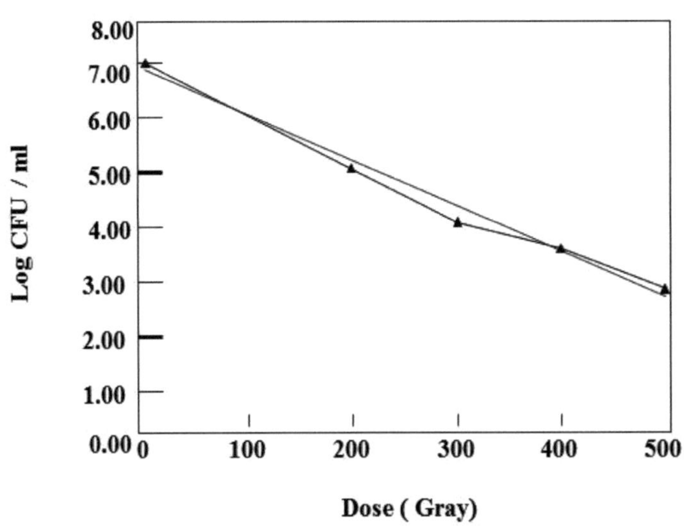


**Figure 6.**
*The oil-degrading bacterial strain (strain 33) survival curve subjected to gamma radiation from a cobalt-60 source.*

| Variable | Treatments | Weeks | | | | | | | | | | |
|---|---|---|---|---|---|---|---|---|---|---|---|---|
| | | **0** | **1** | **2** | **3** | **4** | **5** | **6** | **7** | **8** | **9** | **10** |
| Oil degradation (ppm) | Control | 0 | 0 | 20.6 | 40 | 60 | 100.5 | 152 | 200.6 | 413 | 507 | 510 |
| | | | | f | f | f | f | g | g | g | g | g |
| | Parent strain | 0 | 0 | 100 | 150 | 508 | 270 | 3704 | 6006 | 7790 | 7835 | 9000 |
| | | | | f | g | g | g | g | f | e | e | d |
| | Mutant strain | 0 | 0 | 101 | 1552 | 3420 | 7900 | 11,550 | 11,626 | 15,201 | 15,896 | 16,017 |
| | | | | f | g | e | e | c | c | b | a | a |
| Nitrogen utilization (ppm) | Control | 0 | 0 | 0 | 0 | 0 | 10 | 14.1 | 14 | 14 | 14 | 14 |
| | | | | | | d | d | d | d | d | d | d |
| | Parent strain | 0 | 0 | 13.3 | 28 | 40 | 55.33 | 84.66 | 94 | 112 | 140 | 140 |
| | | | | d | d | d | d | d | d | d | d | d |
| | Mutant strain | 0 | 44 | 55 | 86 | 108 | 423 | 574 | 602 | 610.6 | 614 | 614 |
| | | | d | d | d | d | c | b | a | a | a | a |
| Phosphorus utilization (ppm) | Control | 0 | 6.45 | 19 | 43 | 105 | 206 | 230 | 240 | 247 | 247 | 247 |
| | | | y | x | v | s | r | p | o | n | n | n |
| | Parent strain | 0 | 277 | 58 | 206 | 2257 | 264 | 270.8 | 275.1 | 277 | 299.6 | 280 |
| | | | w | u | q | m | l | k | j | i | d | g |
| | Mutant strain | 0 | 96.66 | 206 | 247 | 264 | 277 | 281 | 288 | 318 | 320 | 325 |
| | | | t | r | n | L | h | k | e | c | b | a |

| Variable | Treatments | Weeks | | | | | | | | | | |
|---|---|---|---|---|---|---|---|---|---|---|---|---|
| | | 0 | 1 | 2 | 3 | 4 | 5 | 6 | 7 | 8 | 9 | 10 |
| Potassium utilization (ppm) | Control | 0 | 16 | 16 | 16 | 55 | 94 | 195 | 211 | 222 | 250 | 250 |
| | | | p | p | p | o | m | j | i | h | g | b |
| | Parent strain | 0 | 16 | 55 | 55 | 90 | 172 | 289 | 280 | 300 | 300 | 300 |
| | | | p | o | o | n | k | e | f | d | d | a |
| | Mutant strain | 0 | 16 | 55 | 94 | 289 | 289 | 320 | 320 | 320 | 327 | 323 |
| | | | p | o | m | e | e | e | e | b | a | a |

**Table 1.**
*The effect of gamma-ray treated soil on the degradation of petroleum hydrocarbons and utilization of other nutrients. The same letters mean those values are not significantly different from each other.*

soil and the soil sample undergoing the experiments was investigated regarding total petroleum hydrocarbons, nitrogen, phosphorus, potassium, moisture, and bacteriological activity regarding solid-in-colony count per unit volume. The results in **Table 1** displayed the effect of gamma-ray-treated soil on the degradation of petroleum hydrocarbons.

The soil collected from the un-augmented bacteria (control) and augmented with the bacteria unirradiated (33 strain), and irradiated bacteria (300-B) microcosms were observed to be blackish-gray at the initial time point (0 hour). Due to bacterial action, the soil color in the control microcosm remained the same; however, the treated microcosm showed a significant decrease in the intensity of soil discoloration.

Furthermore, the soil inoculated with the parent strain also showed a reduction in intensity, but not to the same extent as the treated microcosm. The results of the oil degradation for different enduring intervals indicate that the mutant strain 300-B is the most effective petroleum degrading agent, followed by the parent strain (33) and control at all times. The peak western oil destruction occurred at 10 weeks, with the lowest at 1 week.

### 9.4 Bioremediation of hydrocarbons

In this study, adding the specific irradiated bacteria improved the rate of degradation of hydrocarbons in soil over time. Treatment 3 involves augmentation with the mutant bacterial culture and nutrients. Treatment 1, with no nutrients, is controlled without bacteria, while in Treatment 2, the soil was augmented with unirradiated bacteria.

The best results relative to bioremediation were observed in Treatment 3, which shifted from a total hydrocarbon concentration of 30,000 mg/kg (3%) to 14,000 mg/kg (1.4%), a volumetric reduction of 53.45%. In comparison, the soil with the parental strain and nutrients decreased by only 30% after 75 days.

Hydrocarbon loss was markedly slow in the soil in control, untreated microcosms at 300 mg/kg (0.3%). The rest of the hydrocarbon volume from Treatment 1 came

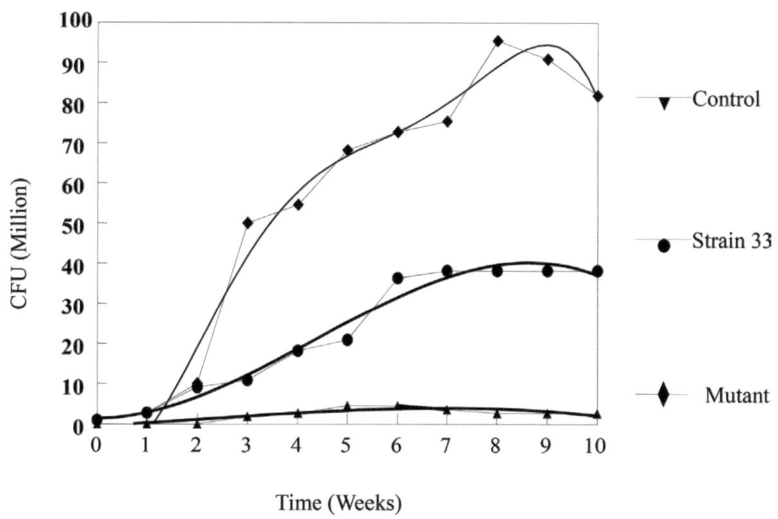


**Figure 7.**
*Presence of microorganism in different soil microcosms for petroleum decontamination.*

solely from the spiking solution. A notable statistical change was also present in the time and type of strain used. The soil was checked for the utilization of essential nutrients such as nitrogen (N), phosphorus (P), and potassium (K). Results suggest that the gamma-ray irradiated bacteria effectively utilized more nutrients, resulting in efficient hydrocarbon degradation.

### 9.5 Counts of microorganisms and colony-forming units (CFU/g)

Upon augmentation, the total viable counts for both the parent and mutant strains were noted as $1.0 \times 10^6$ CFU/g (dry weight) in bacterial augmented microcosms. A marked increase in microbial cell counts indicated that petroleum hydrocarbons could be metabolized and utilized as the sole carbon source. Without bacterial inoculation, a separate microcosm was kept as a control.

Both parent and mutant strain microbial populations remained similar during the first fortnight. However, differences in population enhancement became apparent in week three (**Figure 7**). Between the strains, marked differences in microbial growth were found during the sampling done after 3 weeks, which increased as time passed, resulting in the degradation of petroleum contents in contaminated soil.

## 10. Challenges and future directions

For a long time, petroleum hydrocarbons have been phytotoxic and zootoxic. Contamination by oil necessitates a coordinated effort from scientists and industry policymakers to formulate acceptable solutions to the issue of the petroleum-contaminated media effect. Studies on degradation techniques (specifically, the bioremediation of petroleum hydrocarbons in soil and water) indicate that some fractions (60–90%) of petroleum's constituents are biodegradable.

Encapsulating gamma sources in mobile units and combining them with nanotechnology (e.g., $TiO_2$ nanoparticles for pollutant adsorption) improves efficiency and safety. Strict licensing for radioactive materials delays implementation. Misconceptions about radiation risks necessitate community engagement. Horizontal gene transfer between irradiated and native microbes is mitigated using non-pathogenic strains.

Gene-edited strains can be designed using CRISPR with radiation-inducible degradation pathways: gamma + electrokinetic remediation for clay soils. Machine learning models predict optimal doses (e.g., 3.2 kGy for diesel, 4.5 kGy for crude oil) based on pollutant composition. Synthetic communities combining *Pseudomonas* (alkane degradation) and *Streptomyces* (PAH breakdown) achieve 95% TPH removal.

## 11. Conclusion

As previously mentioned, the synergistic application of gamma irradiation and bioremediation strategies is one of the novel approaches developed to tackle the bioremediation of petroleum pollution. The phenomenon of radionuclide-gamma-ray interference can enhance the metabolic activity of bacteria by mutation-selection, which results in more efficient degradation of complex petroleum hydrocarbon compounds in a petroleum hydrocarbon-contaminated environment. This chapter is dedicated to the radiation-induced pathways of elevating microflora enzymatic degradation processes, such as incorporating functional groups into hydrocarbons that make them accessible to enzymes for cleavage.

Experimental evidence has been obtained that microbes exposed to gamma radiation possess greater tolerance levels of petroleum pollutants and perform more rapid biodegradation. Furthermore, substantial improvements in treating soils and seawater contaminated with petroleum products have been achieved by combining gamma radiation with bioremediation techniques like bioaugmentation and biostimulation.

Even though there are still difficulties, such as optimizing the radiation dose and possibly other secondary effects on the environment, laboratory and field studies provide insights regarding gamma-biological remediation, which is quite promising. The focus must be on constructing mobile modular systems in different environments along with advanced technologies such as nanotechnology and clustered regularly interspaced short palindromic repeats (CRISPR) gene editing for more effective microbial modification. Such developments could enable the creation of inexpensive yet effective strategies to prevent petroleum pollution and promote long-term ecological health.

## Acknowledgements

I am grateful to NIBGE for providing all of the resources and labs for my study, and I also appreciate NIAB for their assistance in supplying the essential equipment.

## Author details

Razia Tahseen[1*], Muhammad Ravish Khan[2] and Muhammad Imran[3]

1 National Institute for Biotechnology and Genetic Engineering (NIBGE), PIEAS, Faisalabad, Pakistan

2 National Institute for Biotechnology and Genetic Engineering (NIBGE-C), PIEAS, Faisalabad, Pakistan

3 Department of Chemistry, Faculty of Science, Research Center for Advanced Materials Science (RCAMS), King Khalid University, Abha, Saudi Arabia

*Address all correspondence to: raziatahseen@gmail.com

## References


[1] Makhlouf B, Abdelsadek Z, Hamada B, Masset PJ. Improvement of the octane number of petroleum naphtha using a molten salt-based thermo-catalytic process. South African Journal of Chemical Engineering. Apr 2025;**52**:170-182. DOI: 10.1016/j.sajce.2025.02.005

[2] Xue J, Dun Y, Wang S, Cheng D, Qiao Y, Liu Y, et al. Enhancement of bioremediation efficacy in petroleum-contaminated marine environments via Quorum sensing: Mechanistic insights and efficacy. Journal of Environmental Chemical Engineering. 2025;**13**(1):115281. DOI: 10.1016/j.jece.2024.115281

[3] Sun Y, Guan J, Yue G, Xiu Y, Wang H, McHugh T, et al. Chemical composition and fingerprints of headspace vapors of 37 petroleum products and non-aqueous phase liquid samples. Environmental Forensics. 2025;**26**:1-16. DOI: 10.1080/15275922.2025.2459405

[4] Rahaman M, Paul S, Das NC. Catalytic and noncatalytic pyrolysis of waste plastics: Comparative analysis of liquid products from polyethylene, polypropylene, and mixed plastics with diesel and petrol. Industrial and Engineering Chemistry Research. 2025;**64**(6):3240-3253. DOI: 10.1021/acs.iecr.4c04155

[5] Satyapal GK, Haque R, Kumar N. Gamma irradiation in modulating arsenic bioremediation potential of *Pseudomonas* sp. AK1 and AK9. International Journal of Radiation Biology. 2024;**100**(6):934-939. DOI: 10.1080/09553002.2024.2345137

[6] dos Santos Oliveira JA, Polli AD, Ferreira AP, Lopes NB, Mangolim CA, Vicentini VEP, et al. Radiotolerant endophytic bacteria and analysis of the effects of 137Cesium on the metabolome of Pantoea sp. Brazilian Journal of Microbiology. 2024;**55**:1-12. DOI: 10.1007/s42770-024-01458-z

[7] Ziyadi S, Iddar A, Kabine M, El Mzibri M, Moutaouakkil A. Changes in growth, morphology, and physiology of *Tetrahymena pyriformis* exposed to continuous Cesium-137 and Cobalt-60 gamma-radiation. Current Microbiology. 2022;**79**(2):61. DOI: 10.1007/s00284-021-02684-6

[8] Chmielewski AG. Future developments in radiation processing. In: Applications of Ionizing Radiation in Materials Processing. 1st ed. Dorodna Warszawa, Poland: Institute of Nuclear Chemistry and Technology; 2017. pp. 501-516. DOI: 10.1007/978-3-319-54716-6_8

[9] Fan X, Niemira BA. Gamma irradiation. In: Electromagnetic Technologies in Food Science. U.S. Department of Agriculture, Agricultural Research Service, Eastern Regional Research Center. Wyndmoor, PA, USA. 2021. pp. 53-73. DOI: 10.1002/9781119759522.ch3

[10] Duan XG, He YM, Li ZQ, Li S. Enhanced biodegradation of oily wastewater through nuclear irradiation mutation and statistical experimental methodology. Journal of Water Reuse and Desalination. 2015;**5**(3):301-311. DOI: 10.2166/wrd.2015.106

[11] Thakur A, Kumar A. Emerging paradigms into bioremediation approaches for nuclear contaminant removal: From challenge to solution. Chemosphere. 2024;**352**:141369. DOI: 10.1016/j.chemosphere.2024.141369

[12] Fathy R, Omara AM. Isolation and optimization of polyphosphate accumulating bacteria for bio-treatment of phosphate from industrial wastewater. Environmental Technology. 2024;**45**(21):4314-4333. DOI: 10.1080/09593330.2023.2248558

[13] Raghu G, Balaji V, Venkateswaran G, Rodrigue A, Maruthi Mohan P. Bioremediation of trace cobalt from simulated spent decontamination solutions of nuclear power reactors using *E. coli* expressing NiCoT genes. Applied Microbiology and Biotechnology. 2008;**81**:571-578. DOI: 10.1007/s00253-008-1741-6

[14] Sarkar S, Ghosh B, Ghosh S. Microbial interventions in bioremediation of nuclear waste. In: Radioactive Pollutant: Sources, Issues and Remediation. Cham: Springer Nature Switzerland; 2024. pp. 325-352. DOI: 10.1007/978-3-031-73796-1_15

[15] Nayak T, Sengupta I, Dhal PK. A new era of radiation resistance bacteria in bioremediation and production of bioactive compounds with therapeutic potential and other aspects: An in-perspective review. Journal of Environmental Radioactivity. 2021;**237**:106696. DOI: 10.1016/j.jenvrad.2021.106696

[16] Khalid SA, Elsherif WM. Types of microorganisms for biodegradation. In: Handbook of Biodegradable Materials. Cham: Springer International Publishing; 2022. pp. 1-27. DOI: 10.1007/978-3-030-83783-9_2-1

[17] McGenity TJ. Introduction to the Isolation and Cultivation of Microbes Involved in the Hydrocarbon Cycle. Berlin Heidelberg: Springer; 2017. pp. 1-25. DOI: 10.1007/8623_2015_177

[18] Alaidaroos BA. Advancing eco-sustainable bioremediation for hydrocarbon contaminants: Challenges and solutions. Processes. 2023;**11**(10):3036. DOI: 10.3390/pr11103036

[19] Agathos SN. Biocatalytic weapons against micropollutants. Environmental Engineering and Management Journal. 2012;**11**:SS99-SS104

[20] Rosato A. Bioremediation enhancement of marine sediments contaminated by crude oil with biogenic pollutants mobilizing agents and biosurfactants [dissertation thesis]. Italy: University in Bologna; 2017. DOI: 10.6092/unibo/amsdottorato/8143

[21] Mayekar PC. Accelerating the Biodegradation of Poly (Lactic Acid) at Mesophilic Temperatures. US: Michigan State University; 2024. DOI: 10.25335/2vjm-h186

[22] Al-Hawash AB, Alkooranee JT, Zhang X, Ma F. Fungal degradation of polycyclic aromatic hydrocarbons. Indian Journal of Pure & Applied Biosciences. 2018;**6**(2):8-24. DOI: 10.18782/2320-7051.6302

[23] Behera S, Das S. Potential and prospects of Actinobacteria in the bioremediation of environmental pollutants: Cellular mechanisms and genetic regulations. Microbiological Research. 2023;**273**:127399. DOI: 10.1016/j.micres.2023.127399

[24] Jaiswal AK, Rho HM. Regulation of genes encoding NAD(P)H: Quinone oxidoreductases. Free Radical Biology and Medicine. 2002;**29**(3-4):254-262. DOI: 10.1016/S0891-5849(00)00306-3

[25] Gao C, Gou N, Yuan S, Lan J. A quantitative toxicogenomics assay reveals the evolution and nature of

toxicity during the transformation of environmental pollutants. Environmental Science and Technology. 2014;**48**(15):8855-8863. DOI: 10.1021/es501222t

[26] Yagnik SM, Arya PS, Raval VH. Microbial enzymes in bioremediation. In: Biotechnology of Microbial Enzymes. Rajkot, Gujarat, India: Department of Microbiology, Christ College; 2023. pp. 685-708. DOI: 10.1016/B978-0-443-19059-9.00010-4

[27] Aminian-Dehkordi J, Rahimi S, Golzar-Ahmadi M, Singh A, Lopez J, Ledesma-Amaro R, et al. Synthetic biology tools for environmental protection. Biotechnology Advances. 2023;**68**:108239. DOI: 10.1016/j.biotechadv.2023.108239

[28] Harirchi S, Rafieyan S, Nojoumi SA, Etemadifar Z. Microbial biodegradation and metagenomics in remediation of environmental pollutants: Enzymes and mechanisms. In: Omics Insights in Environmental Bioremediation. Singapore: Springer Nature Singapore; 2022. pp. 487-514. DOI: 10.1007/978-981-19-4320-1_21

[29] Choi WY, Haggett KD, Dunn NW. Isolation of a cotton wool degrading strain of cellulomonas: Mutants with altered ability to degrade cotton wool. Australian Journal of Biological Sciences. 1978;**31**(5):553-564. DOI: 10.1071/BI9780553

Chapter 7

# Microwave Assisted Pyrolysis (Map) as a Tool to Displace Some Hazardous Plastic Waste

*Marco Frediani, Piero Frediani, Irene Mellone, Gianni Innocenti, Gianpaolo Oteri and Roberto Simoni*


## Abstract

Hazardous plastic wastes are formed as byproducts of several industrial productions, or end-of-life materials such as end life tire (PFU), plastic packaging, contaminated plastic wastes, wind turbine blades containing glass or carbon fibers, or perfluoroalkyl substances (PFAS). All these materials are potentially hazardous to the environment or human health, especially if they are not properly treated. These hazardous waste materials may be disposed of through Microwave Assisted Pyrolysis (MAP), because this process may convert these products into valuable compounds. The waste is completely disposed of, avoiding the formation of a residue. The gas contains hydrogen, carbon monoxide, and light hydrocarbons; the liquid, also called SynOil, may be employed as a substitute for mineral oil, while the solid contains a large amount of carbon and/or glass or carbon fibers. The energy required by the MAP may be obtained from the products formed, so it is a self-sufficient process. The affordability of the process is very good and in some cases the return on investment is calculated at 1.6 years. Our results together with several literature data and our considerations are reported. Furthermore laboratory data and some of the proposed pilot or industrial plants developed up to now are illustrated. In conclusion, MAP converts hazardous waste materials into new products, reducing their impact and obtaining products considered as green materials because their carbon footprint is negative due to their formation from hazardous end-of-life or waste materials.


**Keywords:** microwave assisted pyrolysis, hazardous waste, contaminated waste, plastic, end life tire

## 1. Introduction

### 1.1 Hazardous plastic waste

Hazardous plastic wastes are present as solid, liquid, or gas, and they may be potentially dangerous for human health or the environment. They may be dangerous from a biological, chemical, or physical point of view, and many of them at the same time. They may also be formed as byproducts of several industrial productions, or end-of-life products such as tires, plastic packaging, contaminated plastic waste, wind turbine blades containing glass or carbon fibers, residual plastics from car demolition, and, finally, as perfluoroalkyl substances (PFAS). PFAS are a class of compounds that are very useful for several applications, but they are persistent organic pollutants causing health and environmental problems. All these materials are potentially hazardous to either the environment or human health, especially if they are not properly disposed of.

## 1.2 Microwave assisted pyrolysis (MAP)

### *1.2.1 Microwave heating*

Electromagnetic radiations may be absorbed by some materials and converted into heat, and the efficiency of this conversion is attributed to a property of the material defined as dissipation factor. This property is defined as $\tan\delta = e''/e'$, where e' is the real component of the dielectric permittivity, while e" is the imaginary component of the dielectric permittivity. The dissipation factor is connected with the ability of the material to promote electrons from the conduction bands. Many parameters affect this property, such as the radiation frequency, the type of material, and the temperature [1, 2]. If the temperature at which this phenomenon takes place is increased, the loss factor improves because more electrons are promoted into the conduction bands. The influence of temperature on the loss factor is characteristic for each material, and remarkable differences are observed among different materials [3].

Two classical microwave (MW) frequencies of 2.45 GHz and 915 MHz are commercially employed in MW heating, avoiding interference with other commercial apparatus, even if all MW frequencies may be employed for heating purposes. Experiments reported in this chapter employ the most common 2.45 GHz frequency. The design of the MW oven affects the efficiency of the oven because the electromagnetic field is increased when the in-phase not absorbed radiation is reflected by the internal wall of the oven [4].

Detection of temperature is quite a challenging problem inside an MW oven because each component of the detector is heated, and a heterogeneous temperature profile is supplied. Furthermore, spark discharge may be observed in the oven with a spot heating of the material at a temperature over 1300 K [3], and this is another problem in temperature detection. A classical thermocouple may be employed only if it is placed in an appropriate cavity not reached by MW. Otherwise, the metals of the thermocouple absorb the MW radiation and are heated, so the temperature measured is related to the thermocouple and not to the material. A fiber optic thermometer may be used, but a close contact between the fiber and the solid sample is required, and this contact is not easily realized. Selecting an infrared temperature detector, the reported value is related to the temperature of the vapors inside the oven, while a different temperature is reached by the solid material [5]. Other, more sophisticated methodologies are reported, but they are not commercially available and are hard to realize. The temperature reached in each experiment is reported, but due to the above-reported problems on its determination, usually, it is not employed for interpretation of results. An MW oven (Bi.Elle s.r.l. - Italy), having 4 × 1.5 KW magnetrons with a total MW emission of 6 KW, was mainly employed in our laboratory experiments reported in this chapter. Two set-up equipments were used [3]:
In the first arrangement (called set-up A), a 1000 mL borosilicate Erlenmeyer flask containing the sample was placed inside the oven, while two condensing systems

cooled at 298 and 263 K, respectively, were connected to the flask. A round-bottom flask was employed to collect liquids, while the gas was sent into a gasometer. Another arrangement (called set-up B) was employed, where a fractionating column was present between the oven and the condensing system. Using this arrangement, high-boiling products remained inside the oven for a long time. A 0.2 m length × 0.03 m internal diameter column was employed, and it was filled with glass spheres (0.004 m diameter).

Different ovens were sometimes employed and the characteristics of these ovens were reported where the results are described.

## 2. Results

The most common option to recycle plastic waste is mechanical recycling. This process takes place through the selection of plastic wastes by type and color, followed by washing, drying, pelletizing, and finally printing to obtain new products from recycled plastic. These new products show lower properties than the same products obtained from virgin plastic, so the applications of these recycled plastics may be reduced. Using chemical recycling, monomers may be obtained, and they may be employed to obtain new plastics having the same properties as virgin plastic. These recycled monomers may be also employed for different utilization, such as chemical feedstocks and fuels.

### 2.1 End life Tire (ELT)

Worn tires that are not suitable for use on a vehicle, as well as defected tires—that is, products having deviations from the specifications of the original manufacturer are defined as End Life Tire (ELT). This definition does not include tire-derived products. Every year, around 1.5 billion ELT are produced [6], and their disposal is problematic because they are not biodegradable, require a large amount of space, and take a very long time for degradation when they are disposed of in a rubbish dump. At the same time, a strong attention must be devoted to the storage of this waste because ELT are flammable, and if burned in the air without any control, severe pollution is caused by the release of highly toxic products.

Several technologies are proposed for a suitable and convenient disposal of ELT. A re-manufacturing process for old tires replacing the tread on worn tires with new tread and called "retread (recap or remold)," may be followed. This is one of the most environmentally friendly ways to recycle ELT because it gives an additional life to ELT. However, this process cannot be repeated more than three times, and it is not possible with very old or damaged tires.

The main end use for ELT is the production of ground crumb rubber, employed in the realization of asphalt for highways or as pavement sealers, to obtain a large number of recycled rubber-containing products, as a filler for new tire production, as a feedstock for chemical devulcanization or pyrolysis processes [7].

ELT can also be employed to realize roads or harborside barriers. Finally, whole or chopped ELT may be employed to produce energy in the manufacturing of Portland cement [8, 9], or in other processes where a high amount of energy is required. ELT, in fact, has a high heating value (HHV), close to commercial petroleum coke, but with lower carbon dioxide production if correlated with carbon. In the production of electricity, ELT may be employed to supply the energy for thermoelectric power plants;

however, during the combustion, sulfur-containing compounds present in ELT produce sulfur dioxide, and a depuration apparatus is required to remove this contaminant from the emissions. In 1990, the amount of ELT waiting for disposal in the USA was around 1 billion, but in 2015, storage of ELT strongly decreased to 67 million tires [10] because part of them were employed to produce energy. Tires recycled in the European Union increased from 25 wt% in 1994 to 95 wt% in 2010, even if half of this waste was used for energy production, mostly in cement manufacturing. However, these processes led to recovery of only the energy content of the ELT, while its chemical content was completely lost.

Through thermal treatments in an anoxic atmosphere, such as pyrolysis and devulcanization, ELT was disposed of [11–13] obtaining three useful products: a gas, a liquid, and a char. Several heating technologies may be employed, such as classical heating, internal partial combustion of tires, heat transported by an inert material, microwave heating, laser heating, solar heating, and so on. Different apparatus may be employed depending on the heat source and the time required by the process, affecting the amount and composition of the three compounds formed [14]. Pyrolysis is one of the possible ways to undergo thermal treatment of ELT to generate value-added products. However, a crucial problem for all of these processes is the heat transfer technology because ELTs are very poor heat conductors. MW irradiation, due to its ability to heat quickly and directly any MW-absorbing material, is one of the most promising technologies to carry out pyrolysis. ELT can absorb MW because it contains a high amount of carbon black, able to convert MW into heat. Most of the MAP experiments reported in this chapter were carried out on a laboratory scale using a batch process, using a frequency of 2.45 GHz and a variable energy up to 6 kW. A pilot plant was also used, equipped with an oven having a power of 32 KW. The influence of operating variables on the recovery of the three products formed was tested. Typical MAP products of ELT were a gas containing a large amount of hydrogen, carbon monoxide, and light hydrocarbons; a liquid having a low viscosity (< 2.9 cP), containing hydrocarbons (aliphatic and single ring aromatic); and a solid residue rich in carbon black (up to 92.03 wt% of carbon), iron (5 wt%), and ash (inorganic oxides).

The process does not produce byproducts because all the compounds formed may be easily employed as "re-cycled raw materials." They have an HHV of 34 MJ/kg for solids, 45 MJ/kg for liquid, and 46 MJ/kg for the gas. These products may be considered bio-attributed materials because they are obtained from waste ELT.

MAP, as reported by Undri et al. [3], was run using different MW power and amounts of chopped ELT. Even if the temperature of MAP strongly affects the results of the process, it is not an affordable parameter in MAP, as above mentioned. MAP was run without any flux of an inert gas, with the aim of collecting a gas having a high HHV because it contains only combustible gas. At the same time, the consumption of the inert gas was eliminated. Experiments were interrupted when gas evolution was not further evidenced. A pyrolysis was considered completed if the FT-IR analyses did not evidence the presence of organic materials in the residue, and the hydrogen detected by ultimate analyses in the bio-attributed carbon black (bCB) was lower than 1 wt%.

Results of some representative tests performed using MAP are reported in the following **Table 1**.

The best conditions for a complete pyrolysis of ELT were obtained using an MW power of 3 KW and a reaction time of 32 min, obtaining a solid (43.2 wt% yield, metal wire depleted), a liquid (42.6 wt% yield), and a gas (14.1 wt% yield). The solid was an odorless black powder (bCB), containing fully pyrolyzed material, while not-fully pyrolyzed residue was a sticky, hard-to-handle solid. Hydrogen was always present in

| Run | Tire (g) | MW power (KW) | Reaction time (min) | T (K) | Gas fraction *(Yield wt%) | Liquid fraction* (Yield wt%) | Solid fraction* (Yield wt%) |
|---|---|---|---|---|---|---|---|
| 1 | 208.8 | 6.0 | 14 | 872 | 27.4 | 31.5 | 41.1 |
| 2 | 212.8 | 4.8 | 15 | 814 | 16.3 | 43.1 | 40.6 |
| 3 | 233.3 | 3.0 | 39 | 764 | 14.1 | 42.6 | 43.2 |
| 4 | 252.1 | 1.5 | 100 | 766 | 14.3 | 20.7 | 65.0 |
| 5 | 1501.1 | 3.0 | 70 | 787 | 9.0 | 39.3 | 50.7 |
| 6 | 502.8 | 3.0 | 59 | 744 | 13.4 | 44.0 | 40.6 |
| 7 | 64.1 | 3.0 | 47 | 617 | 22.4 | 30.1 | 47.5 |

**Metal wire depleted.*

**Table 1.**
*MAP of ELT using set-up A [3].*

the solid in low amounts (<1 wt%) due to traces of volatile organic substances (VOCs) present. These VOCs were due to the batch process in the absence of a carrier gas because the small amount of vapors remained in the oven at the end of the experiment and were absorbed by the residue. Anyway, traces of VOCs in the bCB are not a problem for any further application because VOCs are present in commercial petroleum coke (up to 14 wt%) [15], and hydrogen was higher than 1 wt%. In not-completed pyrolysis (runs 4 and 5), appreciable amounts of VOCs were detected in the residue. Using low MW energy or a large amount of ELT the pyrolysis was not completed and a high amount of VOCs was present in the solid, as confirmed by the IR spectrum. Different initial ELT compositions affect the ultimate analysis of char as reported by many authors [16, 17], especially for what concerns the carbon amount. Sulfur-containing compounds were present in the bCB; they were formed from the products used for vulcanization of rubber. The amount of sulfur in bCB was lower than 50 wt% of the initial sulfur present in ELT, and these results are close to those reported for classical pyrolysis. Anyway, several petroleum cokes contain a higher amount of sulfur than bCB, while other petroleum cokes contain a lower amount of sulfur than bCB. The hydrogen content of bCB was not in the range of regular green coke [12, 18]. The composition of bCB (evaluated through ultimate analysis) is intermediate between bitumen and anthracite coal [15]. The use of bCB to produce energy is possible only if the plant is equipped with an apparatus for the depuration of emissions due to the presence of sulfur oxides. The bCB had a low heating value (LHV) higher than ELT and petroleum coke, and the amount of sulfur is lower than many raw petroleum cokes [15]. The solid shows an apparent density of 0.40–0.56 g/mL, higher than exfoliated graphite [19], but the density of a commercial petroleum coke is about 0.8 g/mL [15], and that one of native or synthetic graphite is about 2.5 g/mL. Probably, the grain size of the virgin carbon black (vCB) employed to produce ELT affects the density of bCB obtained after MAP, and the complete pyrolysis avoids the presence of residual rubber in the bCB. The amount of ash in bCB is in the range of 5.88–9.50 wt%, lower than that reported by several authors [20–23], but higher than vCB. These values may be attributed to the residues of iron in bCB after electromagnetic separation or to inorganic compounds present in other fillers employed in the production of tires. The bCB shows higher ash than the value required for vCB (below 0.5 wt%) employed for tire manufacturing [24]. However, the maximum value of ash for their use as carbon black for tire production must be reconsidered, taking into account the suggestions reported by the major firms for tire production, as cited in the following lines. The surface area (BET test) of all samples shows close values, among 45–72 $m^2/g$; however in samples containing a higher amount of VOCs, the lowest surface area was evidenced. Furthermore, the surface area shows a linear correlation with respect to the hydrogen content. A very close total pore volume (VT) was shown in runs 4 and 5: 0.35 mL/g and 0.38 mL/g, respectively. The VT of runs 1 and 3 was higher (0.82–1.67 mL/g), and much higher than the VT of the samples obtained by López et al. (0.3–0.5 mL/g) [21], suggesting different vCB used for tire productions. Runs 4 and 5 showed a mesopore structure (pore in the range 2–50 nm) of the microporosity of bCB, while the analysis of runs 1 and 3 was not run due to the limitations of the available instrumentation. The inorganic materials in the bCB, detected by ICP-MS analysis after magnetic iron wires removal, showed the presence of residual Fe and Zn as the most abundant metals and a lower amount of Cu. Fe was attributed to part of the original metal wire or non-ferromagnetic particles or microparticles of Fe pasted with carbon and not attracted by the magnet. Zn compounds were formed using the additive ZnO used in tire formulation and/or using the brass coating of metal wires. Cu was leaked from the brass coating of the metal wire. Very low amounts of other metal-containing compounds were shown in the carbon black, and their amounts were lower than the limits required by the European Community. The scanning electron microscopy (SEM) and microanalysis (X-EDS) of bCB do not show any significant morphological variation among the samples obtained from a complete pyrolysis. The results do not evidence any influence of MAP on the starting vCB employed as filler, as well as on vCB morphology. A macropore structure (run 1), which was left after pyrolysis of the polymeric matrix, confirmed the granular structure, with grain diameters in the range of 2–50 $\eta$m together with bright spots and large dark areas. X-EDS microanalysis showed differently colored areas having almost the same composition. Large amounts of C, Fe, Zn Cu, together with Mg, Si, O, and S, were evidenced both by microanalysis as well as ICP-MS analysis. Different results were reported by Ucar et al. [22]; they showed a homogenous distribution of particles, attributable to a different pyrolysis procedure and conditions. The metal wire present in the end life tire (ELF) before and after pyrolysis suggests the following consideration: The starting wire was coated with Zn and Cu to obtain a strong adhesion of the polymeric matrix to the wire. After pyrolysis, a cracked surface was shown by residual wire, while a compact material was present in the deep layers. A mild bending of the damaged surface led to removal of the coating. After pyrolysis, Cu was partially removed, and it was present in the bCB. Cu, Zn, and Fe reacted with sulfur-containing compounds present in the tire, giving amorphous compounds and crystalline ZnS. Sulfur was present in higher amounts on the external surface than in the inner layers of the wire, suggesting a sulfur reaction with the brass coating and the iron wire. The brass coating was modified, causing partial leak of Cu. The brass coating and ZnO filler react with sulfur-containing compounds giving different crystalline forms of ZnS. Runs 1, 3, and 4 showed an analogous X-ray pattern, where the strong and broad peak at 25° is due to amorphous carbon, as reported for SEM morphological analyses [25]. Similar crystalline compositions were shown, and the only peaks evidenced by X-ray were due to zinc compounds. Zinc sulfide crystallized in the cubic system (wurtzite, $\alpha$ ZnS), together with a trace of zinc oxide (zincite, ZnO), were the main compounds present, while sphalerite ($\beta$-ZnS) and only traces of wurtzite ($\alpha$-ZnS) and zincite (ZnO) were evidenced in run 5. Some crystalline forms of Fe compounds were present together with the crystalline forms of Zn in minor amounts. Other elements were shown by ICP-MS and SEM microanalysis, even if in trace amounts, such as zinc silicate ($Zn_2SiO_4$), which was evidenced by SEM microanalysis, even if it was not shown by X-ray analysis, as also reported by Darmstadt et al. [26]. On the contrary, silicates were evidenced by López et al. in the bCB. The presence of the crystalline form of ZnS was a big challenge to obtain an indication of the temperature reached in MAP. ZnO is stable below 973 K and it reacted in the course of MAP giving sphalerite ($\beta$- ZnS), stable up to 1273 K, while if the reaction temperature is improved, sphalerite is converted into wurtzite ($\alpha$-ZnS), in agreement with results reported by Darmstadt et al. [26]. In run 5, sphalerite was evidenced, while the other samples showed wurtzite as the main Zn compound. Taking into account the presence of these crystalline Zn compounds, it is possible to confirm a MAP temperature between 973 and 1273 K in run 5, while the other runs (1, 3, and 4) contain almost exclusively wurtzite, that is, a temperature higher than 1273 K is required to obtain wurtzite. In agreement with these data sample 5 obtained with the lowest MW power (P) with respect to the weight of ELT (M) [$P/M^2$: $1.3 \times 10^{-3}$ $W/g^2$], was heated in a range of temperatures between 793 and 1273 K, and this range of temperature is close to the values detected by Undri et al. [3]. The high amount of ELT led to more efficient heat dissipation (MAP was run using 1.5 kg of ELT), and a lower temperature was reached in run 5. The presence of wurtzite and sphalerite in MAP of tire confirms, sometimes as a spot, the high temperatures obtained during MAP [2, 27–29].

The commercial applications of bCB (sometimes called recycled carbon black: rCB) are of great interest [30], because the economicity of the MAP of ELT on an industrial scale depends on the revenue due to the uses of bCB [18, 22]. bCB may be applied for the production of active carbons after demineralization and physical or chemical activation [17–20]. These applications were very recently reported by Zainal et al. [31] in a review. The carbon black from pyrolysis of ELT (bCB) usually has a higher content of ash if correlated to virgin carbon black (vCB), depending on the kind of ELT used for MAP. A very important application of bCB, as reported by Contec [32], is the production of new tires due to the following considerations:

- bCB can provide 100 wt% replacement for rubber and nonrubber applications of vCB;
- bCB has fewer polycyclic aromatic hydrocarbons (PAHs) compared to vCB;
- Local and regional collection and recycling of ELT can provide an assured source of bCB;
- Prices of bCB are stable, as it does not depend on fossil hydrocarbons;
- Prices for bCB are lower than vCB;
- Pioneer tire manufacturing businesses can enjoy a head start from bCB use;
- The carbon footprint was reduced by 80–90%. For example, the production of bCB requires an estimated formation of 439 kg $CO_2$ per ton, while 2500 Kg $CO_2$ per ton of vCB are produced [25];
- Transition from vCB toward bCB meets the requirements for a circular economy, the EU Green Deal requirements, and the United Nations Sustainable Development Goals [33].

Furthermore, the following considerations are reported by many companies:

- Volvo announced in 2016 [34] "Volvo Car Corporation (Volvo Cars) has approved the use of Enviro's *recycled carbon black (rCB)* chassis plugs that AnVa Polytech supplies to the car manufacturer."

- Bridgestone and Michelin, two of the major leaders in tire production, proposed a Shared Vision for safer, cleaner, and more sustainable mobility, and this involves the use of bCB.

- Volvo's chassis plugs were planned to be realized from rubber containing 100% "recycled carbon black" and the idea was confirmed in 2021 with the announcement "Volvo has provided details of what it believes its ultra-green electric cars will be made from by around 2030; The bumper panels are constructed from a material that consists of fibers made from linseed plants; Interior parts are recycled plastic, seats use 'responsibly-sourced wool' and they are made from old rubber" [35]

- Bridgestone and Michelin estimated a potential market for up to 1.000.000 tons of bCB in 2030 [36].

- Pirelli announced the production of the tire P Zero E, devoted mainly to electric cars, obtained using a 55 wt% mix of biobased and recycled materials as certified by Bureau Veritas. The biobased materials (29–31 wt%) will contain natural rubber, reinforcing textiles, chemicals, biobased resins, and lignin, while recycled materials will be obtained from metallic reinforcement, chemicals, silica, and bCB. In this way, the emission of $CO_2$ equivalent will be reduced by 24 wt% compared to the previous tire produced. The use of virgin material having fossil origins will be reduced by 3 Kg for each tire [37].

- bCB obtained by Tyrebirth from MAP of ELT has the same performance as vCB for the production of new tires.

The gas formed during the MAP of ELT may be employed for the production of electricity, and this energy is sufficient for the needs of the plant.

Recently, Shao et al. [38] reported a study of the process parameters of MAP with the aim of improving the performance of the liquid fraction obtained. They also suggested the utilization of catalysts such as ZSM-5 and SAPO-34 to improve the amount of olefins and cycloolefins, with the best results using SAPO-34. However, the catalyst must be separated and recycled at the end of the process to allow for the utilization of the bCB formed and to avoid the improved cost due to the catalyst supply, so an evaluation between cost and benefit must be taken into account.

The liquid formed during MAP of ELT (sometimes called "SynOil" [39]) has recently improved its interest as a substitute for virgin oil. There are many companies interested in employing SynOil from different sources:

- BASF and Braven Environmental signed a long-term agreement for the use of advanced recycled feedstock [40];

- LyondelBasell claims the production of resin called "Circulen Revive" in Wesseling (Germany) using SynOil obtained from end-of-life plastics, through cracking to ethylene and propylene, and their polymerization to polyethylene and polypropylene. These polymers were bio-attributed through a mass balance certified by International Sustainability & Carbon Certification (ISCC) plus [41];

- BP and Saudi Basic Industries Corporation (SABIC) announced the production in Gelsenkirchen (Germany) of circular plastic called TrueCircle from SynOil. This feedstock, obtained by pyrolysis of end-of-life plastics, was employed to produce olefin and through polymerization the new plastic [42];

- OMV has received support of 81.6 million € from the European Community to develop an industrial plant with a planned production of 200,000 tons/year of SynOil, using its ReOil technology. This production is planned to be used for the production of polyolefin, chemical intermediates, or fuel [43];

- DOW, together with ING, Invest-NL, Polestar Capital, and Vopak, was involved in supporting Xycle in the realization of a new plant in Rottherdam's port to obtain a SynOil through pyrolysis of different plastics [44];

- FreePoint Eco-System (Ohio), a new firm, obtained PyOil (a synonymous of SynOil) in the new plant in Hebron, Ohio (USA), from end plastics waste, and they send it using railway, to Shell USA's refining facility in Norco, Louisiana [45];

- Bridgestone start the tests, as announced at WasteExpo 2025 [46], on a new green tire containing 33% of recycled materials and 37% of biobased material [47], for a total of 70% of green products, as certified by ISCC plus.

Very recently, a review on the thermochemical utilization of ELT has been published by Wang et al. [48] with the aim of promoting an efficient utilization of ELT and environmental protection.

### 2.2 MAP of plastic waste

The global plastic production was $413.8*10^6$ tons in 2023, as reported in **Table 2** [49], and consequently, the amount of end-of-life plastic waste is continuously growing.

At the same time, the alliance to end plastic estimated $360*10^6$ tons of waste plastic, deriving 50% from packaging and 30% from construction, industry, and agriculture. In 2023, 70% of plastic waste remained uncollected; it was dumped into landfills or burned. Europe was the leader in recycling, with a rate of 15%, followed by Asia with 12%, and North America with 5% [50]. Half of the plastic waste collected for recycling by the EU member states (about $3*10^6$ tons in 2011) was exported to Asia, mostly to China, but recently, China has strongly reduced the import of waste plastics. In the UK, the household plastics collected in 2015 were 500,000 tons. These statistics show an improvement in the waste plastic collection worldwide, without any decrease in mixed, nonrecyclable plastic waste.

One of the possibilities to dispose of plastic waste was thermal treatments, where all kinds of plastic may be disposed of. Among these treatments, MAP of end-of-life plastic materials has been proposed and reviewed many times in the last 10 years [51, 52]. These plastics show low efficiency in converting microwave radiation into heat, but in the presence of an MW absorber, sometimes called "catalyst," efficient MAP

| Type[#] | World (413.8*10$^6$ ton) (%) | Europe (54.0*10$^6$ ton) (%) |
|---|---|---|
| PP | 19.0 | 15.7 |
| LDPE | 14.0 | 13.1 |
| PVC | 12.8 | 8.4 |
| HDPE | 12.2 | 8.3 |
| PET | 6.2 | 4.7 |
| PUR | 5.3 | 5.5 |
| PS | 5.2 | 5.2 |
| Thermosetting plastic | 8.6 | 7.6 |
| Other thermoplastic | 7.1 | 10.9 |
| Mechanical recycled | 8.7 | 19.0* |
| Biobased and bio-attributed plastic | 0.7 | 1.4 |
| Chemically recycled | 0.1 | 0.2 |

*[#]PP: Polyopropylene; LDPE: Low density polyethylene; PVC: Polyvinylchloride; HDPE: High density polyethylene; PET: Poly(ethylene therephtalate); PUR: Polyurethane; PS: Polystyrene;*
**As the sum of pre- (5.8%) and postconsumer (13.2%) plastic.*

**Table 2.**
*Plastic production in 2023.*

may be realized. The main employed MW absorbers are silicon carbide, carbon, iron, and so on, even if other absorbers—each one with specific features—are reported in the literature.

The liquid formed during MAP (usually called "SynOil" [39]) has recently improved its interest as a substitute for virgin oil. The technical and economic feasibility of the use of SynOil from the pyrolysis of plastic waste was evaluated in a specific project supported by the European Community studying a 40,000 tons/year plant located in Belgium. The results are reported in **Table 3**.

These data are a strong indication that a 40,000 ton/year SynOil plant is financially attractive for private investors due to the following parameters: (a) the economics of the plant have an IRR over 20%, taking into account oil prices evaluated in August 2015; (b) waste plastics are largely available; and (c) all products obtained from SynOil are requested by the market. Six main variables affect the economic viability of a SynOil plant: (i) the available amount and quality of plastic waste, (ii) the costs of waste plastics, (iii) the cost of capital and operation, (iv) the income from the sale of the products, (v) the tipping fees, and (vi) the location of the plant

| | Code | Unity | |
|---|---|---|---|
| Plant capacity | | ton/y | 40.000 |
| Capital cost | | € | 20.8*10$^6$–27.0*10$^6$ € |
| Operating expenditure | OPEX | € | 2.9*10$^6$ € |
| International rate of return | IRR | % | 20 |
| Net present value | NPV | € | 1.47*10$^6$ |

**Table 3.**
*Results of the SynOil project.*

with respect to the collection of waste plastic to reduce transportation. For instance, the sorting plant of a municipal waste plastic collection, the pyrolysis plant, and the SynOil refining plant are close to one another.

Finally, Kima et al. [53] report the use of SynOil obtained by pyrolysis of waste plastic for the synthesis of olefins, the raw material for the production of plastics.

### *2.2.1 MAP of single or mixed plastics*

Plastic, especially polyolefins, are transparent to MW radiation, so if these materials are irradiated by an MW, they do not efficiently convert the radiation into heat. A MW absorber, such as carbon, iron, SiC, or another absorber mixed with the plastic, is required to perform MAP. In this way, the MW absorber is heated, and the heat is transferred to plastic materials in contact with the MW absorber. The MAP of single or mixed plastic has been realized, and representative results are reported in **Table 4** [54–58].

### *2.2.2 Municipal solid waste (MSW)*

Trash or garbage in the United States and rubbish in Britain is the common name for municipal solid waste (MSW), and it is an everyday item discarded by the public. In the European Waste Catalog [59], MSW is defined as "mixed municipal waste" and coded as 20 03 01. MSW may be derived from many sources; its composition may change with the location where it is collected and during the course of the year. When significant recycling activity is absent, the composition of MSW in developed countries includes waste derived from residue of food, market, garden, plastic, and packaging materials, together with other solid wastes deriving from residential, commercial, institutional, and industrial sources. In a well-developed country having a good waste recycling system, the main components of MSW are plastic films and nonrecyclable packaging materials. Usually, classical MSW does not include waste

| **Polymers*** | **MW absorber** | **Polymer/adsorber (Ratio)** | **Gas** | **Liquid** | **Solid** | **Refs.** |
|---|---|---|---|---|---|---|
| HDPE | Carbon | 1 | 15.7 | 83.9 | 0.4 | [52] |
| PP | Carbon | 1 | 13.3 | 74.7 | 12.0 | [52] |
| PS | Carbon | 1 | 6.7 | 92.3 | 1.0 | [39] |
| PS | Carbon | 1 | 3.0 | 96.1 | 0.9 | [53] |
| PLA | Carbon | 1 | 48.9 | 48.4 | 2.7 | [54] |
| ELT | // | // | 10.3 | 39.2 | 50.5 | [55] |
| PET | ELT | 2 | 25.4 | 26.3 | 48.3 | [55] |
| PE (2.25)/PP(2.5)/ PET(1) | ELT | 1 | 15.3 | 49.7 | 35.0 | [55] |
| PVC | ELT | 2 | 27.1 | 19.7 | 53.2 | [55] |

**HDPE: High density polyethylene; PP: Polypropylene; PS: Polystyrene; PLA: Poly(lactic acid); ELT: End life tire; PET: Polyethylene terephatalate; PE: Polyethylene; PVC: Polyvinyl chloride.*

**Table 4.**
*MAP of single or mixed plastics.*

from industry, agricultural production, medical activity, radioactive compounds, and sewage sludge [60].

MSW may be employed for the production of methane *via* fermentation, a potent greenhouse gas but useful for the production of energy. MSW may also be employed for the production of energy in a cleaner and economical way, such as direct combustion, gasification, plasma arc gasification [61, 62], pyrolysis, and landfill gas capture. Among these technologies, MAP may also be employed to convert MSW into interesting products, taking into account the possibility to convert both plastic materials and biomass present in MSW [63]. As previously reported, plastics are easily converted into the three classical products of pyrolysis, and they may be employed in chemical synthesis and classified as bio-attributed compounds by mass balance, or used for energy production. As reported in several papers, biomass eventually present in MSW is converted into bio-gas, bio-oil, and bio-char, having several applications: For instance, in a bio-refinery, it is possible to separate interesting products, while the residue may be employed for energy production. The process is very interesting because harmful chemicals, such as PFAS, or biological contamination due to pathogens or pollutants, are completely destroyed in the course of MAP, taking into account the results reported in the following paragraphs. Furthermore, the presence of microplastics in MSW, as reported by Ma et al. [64] will be completely eliminated facilitating an environmentally friendly disposal process. Wei et al. [65] very recently reported a strong improvement in gas production using a catalytic MAP technology, has been reported, even if in this way a Ni catalyst is required (15%), and it remains in the solid or must be separated at the end of the process to employ the biochar obtained.

Tannery sludge (TS) eventually present in MSW may be efficiently and safely removed through MAP as reported by Du et al. [66]. In the condition adopted by these authors, there is a weight reduction up to 43% with toxic organic materials transformed into gas and liquid compounds. At the same time, toxic inorganic materials such as Cr(VI) and Cr(III) compounds were converted into $FeCr_2O_4$ and $Cr_2O_3$, that is, they were immobilized in the residue.

### 2.3 Contaminated plastic waste

Contaminated plastic waste may contain harmful chemicals, pollutants, or pathogens, such as Covid-19 masks, posing significant environmental and health risks, impacting ecosystems, wildlife, and potentially human health. These products cannot be treated using mechanical recycling, but they may be successfully treated using MAP as an environmentally friendly alternative to dispose of them and to close the circular economy.

The mechanical recycling of contaminated, as well as mixed plastic waste, cannot be followed for economic and technical problems, while MAP, in the presence of an MW absorber, may be run with particular advantage. It is an environmentally friendly process because it allows for transforming plastic waste into useful products with the formation of a minimum/absent amount of residue, closing the circle of the economy of plastic.

Contaminated and/or mixed plastic waste has been pyrolyzed, respecting the principles of green chemistry, particularly the atom economy (Principle 2) and the design for energy efficiency (Principle 6) [67]. Because all atoms are present in the products of pyrolysis, Principle 2 is completely satisfied: The products (gas, oil, and char) may be reused to produce useful chemicals. Among thermal treatments, MAP may be successfully employed due to the efficient way to transfer the energy supplied from the microwave generators to the MW absorber (carbon), and then to the contaminated plastic. The heat transfer efficiency is very high because it takes place through direct contact between the MW absorber and plastic avoiding any dispersion.

MAP usually works in an anoxic atmosphere, that is, in the absence of oxygen or the presence of a low amount of oxygen, working between 673 and 873 K, forming three products: a gas, an oil, and a char through chemical reactions. Methane, carbon monoxide, light hydrocarbons, and hydrogen are the main components of the gas; hydrocarbons are the main components of the oil, and they can be used as a fuel or as sources of chemicals; a carbon-rich solid material is the prevalent compound in the char, and it may be employed as a precursor for the production of activated carbon. Taking into account the temperature reached in the process, the contamination and the plastic are rapidly destroyed. The energy required for the process may be obtained using the gas products formed in the process itself, because the gas may be efficiently converted, using a turbine, into electric energy. This energy is supplied to magnetrons and converted into MW to perform the MAP.

The E-factor (mass of residual waste formed with respect to the mass of product treated), connected with green chemistry requirements, is completely satisfied by MAP because this process takes place without or with a minimal amount of final waste.

Recent studies have shown the low environmental impact of plastic pyrolysis through a study on the life cycle assessment (LCA).

The thermochemical recycling of contaminated face masks (FM) shows great potential for harmless treatment, rapid quantity reduction, and resource utilization as reported by the following results:

1. Techwave, using MAP and carbon as microwave absorber has pyrolyzed facemasks (FM) used in the course of the COVID-19 pandemic [68]. The facemasks were converted into the classical three products of a MAP [liquid (75%), solid (12%), and gas (12%)]. The liquid showed the same composition as that obtained from pyrolysis of PP, the solid contains carbon as the main component (>98%), and the gas contains hydrogen, CO, and light hydrocarbons. Products were checked for any possible residual contamination, but, as expected, contaminants and/or pathogens were absent, as expected, taking into account the conditions of the MAP.

2. A dedicated pyrolysis plant was employed by the Fraunhofer Institute UMSICHT, in cooperation with SABIC and Procter & Gamble (P&G) to dispose of used facemasks (it is not specified what kind of heating was employed) [69]. The masks were automatically shredded and subsequently pyrolyzed, obtaining a pyrolysis oil. Analysis of this product did not show any residual pollutants or pathogens, such as the Coronavirus. SABIC employed the pyrolysis oil to produce new plastic, obtaining a high-quality PP.

3. The environmental waste international (EWI) [70] reports the ability to use the proprietary MAP plant to eliminate any contamination and toxic/infectious compounds in materials coming from hospitals and clinics. Furthermore, EWI affirms its ability to safely eliminate any pathogens present in the contaminated waste, avoiding any risk for the public. The plants were carefully designed with the aim of strongly reducing any handling of the materials and completely sterilizing every effluent from the contaminated wastes.

## 2.4 Per- and polyfluoroalkyl substances (PFAS)

In synthetic organo-fluorine compounds, there are many C-F bonds (the strongest bonds in chemistry) and some oxygen (O), hydrogen (H), nitrogen (N), and sulfur (S) containing functional groups. These compounds are called "Per- and polyfluoroalkyl substances (PFAS)," [71, 72] they are very stable and they are sometimes defined as "forever chemicals" [73, 74]. In consideration of their stability, they remain in the environment, moving through soil, groundwater, and surface water. Contamination of drinking water by PFAS is reported, as well as their bioaccumulation in fish, wildlife, and humans. A possible ecotoxicity and human toxicity of PFAS is reported because their presence was evidenced in the liver, thyroid, and reproductive systems. PubChem reports around 7 million products having these characteristics [75]. The first PFAS was Teflon (Polytetrafluoroethylene (PTFE) reported in 1938 by Chemours, a spin-off from DuPont), having interesting performances such as a very high hydrorepellency and nonsticking properties; furthermore, it resists heat, oil, grease, water, and practically it is chemically inert. Several fluorine-containing compounds showing properties very close to Teflon, and reported as PFAS, are synthesized in the course of years and are presently commercially available. In consideration of the characteristics of PFAS, these products are traditionally used in fabrics and industrial materials to improve the nonstick and oil/water repellent properties of carpets, food packaging, Teflon coatings, and fire-fighting foams coating [76–78]. Furthermore, several cosmetics and personal care products, among which nail polish, foundation, concealer, lip balm, lipstick, eyeliner, and mascara [79, 80] contain PFAS.

The United States Environmental Protection Agency (USEPA), in consideration of the above-reported properties of PFAS, imposes the maximum contaminant level (MCL) of oxygen-containing PFAS (PFOA) at 4.0 ppt and PFOS (oxygen and sulfur-containing PFAS) at 4.0 ppt, and sets a goal of the complete elimination of them, as well as the other PFAS compounds [81] in the future. Likewise, the European Union (EU) targeted a limit value of 500 ng/L in 2021 for each PFAS in drinking water [82].

PFAS are products very persistent in the environment and in the human body and very difficult to eliminate, due to their characteristics. The main way to remove PFAS is through absorption on activated carbon or reverse osmosis if they are present in water. PFAS are removed by washing when they are present in the soil. However, these methods transfer the PFAS from one medium (water or soil) into another medium (carbon or water or membrane), but they do not eliminate PFAS. PFAS are also present in sewage sludge.

Thermal treatments, such as combustion, pyrolysis, gasification, and hydrothermal liquefaction, were considered as emerging technologies to destroy PFAS compounds. Bamdad et al. [83], using a high-temperature pyrolysis process, were able to destroy the chemical structures of PFAS, breaking the C-C and C-F bonds. Incineration, carried out at temperatures ranging between 1700 and 2300 K destroys the PFOA and PFOS adsorbed by granular activated carbon (GAC), but dangerous emissions with possible long lifetime effects, such as dioxins and furans, were formed, ruling out this process among those possible to follow for the elimination of PFAS. Pyrolysis was tested for the thermochemical upgrade of PFAS-containing waste biomass, forming a gas, a liquid, and a solid product. The feedstock was heated at temperatures between 800 and 1100 K in an anoxic atmosphere or in the presence of a low amount of oxygen. Bamdad et al. tested the possibility of eliminating PFAS eventually present in carbon-containing PFAS, demonstrating the destruction of PFAS from biosolids using a pyrolysis process carried out at elevated temperature (HTP). A -continuous bench scale HTP system, using a mixed-bed reactor, was able to process three types of PFAS-containing biosolids. The samples were obtained from water recovery sites across the US, and the process was carried out at temperatures between 780 and 1000 K. Working at 1000 K, PFAS were completely eliminated from the solid residue, and only 10% of them remained in the gas as low molecular weight PFAS, that is, low-toxic compounds. The fate of the other PFAS was not reported; they were presumably converted into inorganic compounds.

Hušek et al. [84] reported the possibility of pyrolyzing PFAS-containing sewage sludge. They prepared and pyrolyzed an artificial PFAS-contaminated sewage sludge. It was necessary to work at T > 700 K to remove PFAS and other organofluorine compounds. The results were confirmed by employing a commercial sludge in a pyrolysis unit at the Bohuslavice-Trutnov waste water treatment plant (WWTP) (Poland) working at a temperature between 500 and 900 K to evaluate the lower temperature required for effective PFASs elimination. Working at T > 1000 K, PFAS was not only cleaved but also mineralized.

Taking into account the results reported by Bambdad et al. and by Hušek et al., it is possible to propose the elimination of PFAS using the MAP process. This technology, in fact, works with a very efficient heat transfer at temperatures largely higher than 700 K [25], and in these conditions, the PFAS will be completely destroyed and mineralized.

## 2.5 End life glass or carbon fibers

### *2.5.1 End life glass fibers (ELGF)*

Today, glass-reinforced fibers (GF) are commonly employed in industries such as aerospace, automotive, marine, and construction. Furthermore, ELGF use is improved due to the diffusion of wind turbines for the production of electric energy. Wind turbine blades are predominantly composed of glass-reinforced fiber composites, and at the end of their use, they must be disposed of. MAP as a method to recycle ELGF was reported by Zhang et al. [85] using SiC as adsorbent and recovering 84–88% of solid material, 5% of tar, and 17% of gas. Also, Akesson et al. [86] recovered glass fiber through MAP, obtaining 70% of the starting composite, while the glass fiber preserves 75% of its tenacity. These fibers may be employed to prepare new GF composites, where the amount of recycled fiber was 25%, as well as polypropylene reinforced glass fiber composites [87].

An overview is recently reported by Xu et al. [88] on the methods suggested to dispose and recover glass fibers from end life wind turbine blades (ELWTB-GF). Among the several processes proposed, there is pyrolysis, with special attention to the possibility to recycle the glass fibers and utilizing these materials for a second life cycle. The review examines the possible degradation of recovered glass fibers (rGF) in the course of the processes employed. Obviously, there is a correlation between the characteristics of the rGF and their future potential application, because the economic viability of the recycling method is strongly affected by the sustainable application of rGF. Several suggestions on the direction of future research are reported, as well as technological advancements needed in this field. Finally, Jasinska et al. [89] report a review on recycling methods of wind turbine blades (WTB) and they analyze the application of rGF in cementitious composites.

### *2.5.2 End life carbon fiber (ELCF)*

Carbon fibers (CF) are employed in all fields of glass ones, that is, in aerospace, automotive, marine, and construction industries, with a lot of advantages because they are lighter, stronger, safer, and offer a higher aesthetic value, even if the main drawback is the higher cost. However, the durability and performances of carbon fibers often overcome material costs. One of the fields where carbon fibers were employed is the production of wind turbine blades. As reported for glass fibers, it is a necessity to recover carbon fibers and resin from wind turbine blades (WTB) at the end of their life cycle. Recycling of unsaturated polyester resin (UPR) containing ELCF has been reported by many authors, and pyrolysis is considered the main process for their recycling [90]. As expected, in the course of the recovery process, the properties of carbon fibers (rCF) were partially lost. Presumably, thermal damage is the main cause and the loss of performances was lower using thermal heating than microwave heating. This result suggests an inadequate temperature monitoring during microwave pyrolysis, causing a fiber overheating and, consequently, a thermal damage. In the development of the process, control of process parameters must be improved to overcome the damage of rCF. However, MAP of ELCF offers attractive areas of application where the mechanical properties of the recovered fibers are not the main characteristics, taking into account the short processing time. In a few minutes, ELCF may be quickly reduced, and rCF are employed as fillers or additives for plastics. The aim of this utilization is an improvement of the electrical conductivity of the composites rather than the mechanical properties of the fiber. An improvement of this method was recently reported using a high-power laser radiation to remove the resin matrix without damaging the rCF [91]. Yousef et al. [92], with the aim to obtain energy and rCF by decomposition of the resin matrix in ELCF, test a pyrolysis process using 250 g of ELCF at 900 K. The resin ligand was decomposed, forming a liquid and a gas, together with a rCF-containing residue. An analogous process may be realized using MAP as a heating technology.

The catalyzed MAP of ELCF from WTB was reported by Yao et al. [93] with special attention to the products formed by the pyrolysis of epoxy-resins. They obtained phenols as the main products working at 723 K and using the zeolite ß30, a phenol yield of 36% was reported.

Chin et al. [94] study at the MAP of ELCF with the aim of analyzing the gas formed and the mechanical properties of rCF. The MAP was carried out at 1220 K, and the gas, as expected, contains CO, $CO_2$, and particulates. The crystallite size of rCF was reduced from 0.297 to 0.2333 Å while the tensile strength and tensile modulus were improved. Finally, Dega et al. [95] report a maximum resin elimination of 80% and they obtain a 100% elimination of the resin if the MAP is followed by a chemical dissolution.

A CF pre-impregnated with a resin, typically an epoxy resin, and the proper curing agent (called prepreg-CF), is an intermediate material in the form of sheets, ready to obtain final reinforced CF materials. A large amount of waste is formed in the working process to obtain the final products. A possible way to dispose of the waste prepreg-CF was reported by Hao et al. [96] through a MAP obtaining rCF having a reduced strength (lower than 20%). Controlling the temperature of the process, they obtain a lower reduction of strength while the profile of the surface remains clean, soft, and almost the same functional groups are present.

Li et al. [97] report the MAP of ELCF and the use of rCF to improve the performance of concrete. The specimens obtained using 10% of rCF showed an improvement of impact resistance over 23 times that of the benchmark specimen, while other mechanical properties, such as compressive, flexural, and splitting tensile strength, were improved by 14.2, 565.5, and 22.5%, respectively.

### 2.6 Waste electric and electronic equipment (WEEE)

Waste electric and electronic equipment (WEEE) are the residue of electric and electronic apparatus. These products should not be disposed of among MSW, but they must be sent to a specialized waste collection, where they are separated and recycled. Taking into account the diffusion of electric and electronic apparatuses, the amount of WEEE was strongly improved. Today, a part of this residue cannot be recycled and is sent to a landfill. A possible way to reduce the amount of WEEE is the MAP, obtaining a gas, an oil, and a residue. Six different solid fractions of WEEE (from light dust to particles sized 7–12 mm) were tested by Anderson et al. [98], showing a reduction of the residue correlated with the reaction time, while the temperature of the process does not show any influence.

MAPs of the end-of-life computer cases were run by Rosi et al., using different MW absorbers and setups [99], obtaining a liquid fraction in a large amount (yield 76.6 wt%) containing a lot of aromatic compounds (styrene, xylenes), while halogenated compounds present in the starting WEEE, were completely destroyed. The solid was strongly reduced (yield of solid 14.2 wt%), while the gas fraction was 9.2%.

Another fraction of WEEE, formed by waste printed circuit boards (WPCB), was disposed of by Suriapparao et al. through MAP using graphite or activated carbon as MW absorber [100]. A high amount of residue was formed (yield 58 wt%), and the amount of MW absorber tar and gas was reduced. A significant amount of phenolic compounds and phenylphosphates were present in the solid residue, while $CO_2$ was the main component of the gas. Using activated carbon as the MW absorber, the concentration of $H_2$ in the gas was improved. The residue contains a large amount of Cu, Pb, Ti, and a minor amount of Bi, Fe, and Ca.

MAP of WPCB was reported by Liu et al. [101], using a flux of nitrogen (0.2 lt/min), obtaining, at 873 K, a solid (75%), a liquid (15%), and a gas (10%). The breaking of the C-Br and C-O bonds in brominated epoxy resins was promoted, and $C_6$–$C_9$ compounds, among which phenols and 4-isopropylphenol are the main components. The residue contains Cu (22.8%), Sn (0.84%), Ag (0.025%), and bromine-containing compounds (2.8%).

A very large conversion of bromine-containing compounds, present in WEEE as flame retardants, was reached by Zhang et al., working at 1023 K [102], and the conversion was increased by running at higher MAP temperature. If basic compounds such as $K_2CO_3$, $Na_2CO_3$, and NaOH were mixed with the WEEE, bromide compounds (KBr and NaBr) were present in the residue, while the amount of HBr in the gas was strongly reduced. The conversion was improved with an increase in the time of the pyrolysis. Phenolic compounds were the main compounds present in the liquid fraction (91.1%) when the temperature of the MAP was 623 K, while polycyclic and monocyclic aromatic hydrocarbons (except phenols) were improved to 20.5% and 19.0%, respectively, when the MAP was carried out at an higher temperature (923 K). $CO_2$, CO, $CH_4$, and $H_2$ were present in the non-condensable gas fraction, and these compounds were significantly increased if a higher temperature was employed to carry out the process. If the MAP was run in the presence of a catalyst (ZSM-5 or kaolin), the liquid contains a large amount of monocyclic aromatic hydrocarbons (except phenols) and $C_{11}$–$C_{20}$ compounds. On the contrary, the amount of the gas formed was reduced.

A review about the pyrolysis process devoted to the recycling of plastic components present in the WEEE was reported. Special attention was paid to the liquid fraction (monomers, hydrocarbons, phenols, etc.) [103] and to the elimination of bromine-containing compounds in the WEEE through solvent extraction, supercritical fluid technology, and so on.

Finally, the waste from end-of-life solar cells and photovoltaic apparatus must be considered as WEEE. MAP and other treatments of waste crystalline-silicon solar modules were reported by Huang et al. [104], obtaining a recovery of Al (94.9%), Ag (95.7%), and Si (83.0%). Other metals such as Cu, Ga, In, Se were recovered from waste solar cells using a MAP followed by thermal processes [105].

### 2.7 Car fluff

Residual crushing of vehicles, usually called "car fluff," is the residue of vehicle scraps. This waste includes seals, rubbers, textiles, plastics, and tire fragments from end-of-life vehicles. In the demolition of end-of-life cars, the metal components were initially separated together with metals strongly linked to plastic parts. The residue of this crushing is usually called "car fluff," and it includes seals, rubbers, textiles, plastics, and tire fragments from end-of-life vehicles. The amount of this residue is around 25% of the starting weight of the car.

Camoter International, for instance, sells apparatuses devoted to shredding and separation of car fragments, as well as extruders to obtain briquettes from the "car fluff." These briquettes may be employed as a source of energy because they have a high HHV, but in this way, only the heating value of the briquette was recovered. Taking into account the results of MAP of mixed polymeric compounds reported in **Table 2**, "car fluff" has been pyrolyzed through the MAP, using carbon as MW absorber, obtaining the classical three fractions of a pyrolysis process. These fractions have a composition close to those obtained from mixed plastics, that is, a gas rich in hydrogen and light hydrocarbons, a liquid containing aliphatic and aromatic hydrocarbons, and a residue containing high amounts of carbon [106].

## 3. From laboratory experiments toward industrial plants

MAP was carried out in many laboratory tests, as reported in the literature, using a household MW-modified oven with an electric power absorption of 1 KW and supplying a MW power of 0.8 KW, or different MW ovens specifically realized for laboratory use, such as that one above reported by Frediani et al. [3]. This last oven was realized by Bi.Elle (Italy) and absorbs 8 KW of electric energy, giving an MW power of 6 KW. Another oven was employed by Tyrebirth (Italy), using electric energy of 4 KW, able to supply an MW of 3.2 KW. Other MW ovens were employed by Environmental Waste Technology, Canada (EWI) [75], Pyrowave (Canada) [107], Molectra (Australia) [108], and so on.

One pilot plant was realized by Tyrebirth using an oven fed by 32 generators [109], each of them absorbing an electric energy of 1 KW and able to supply 0.8 MW.

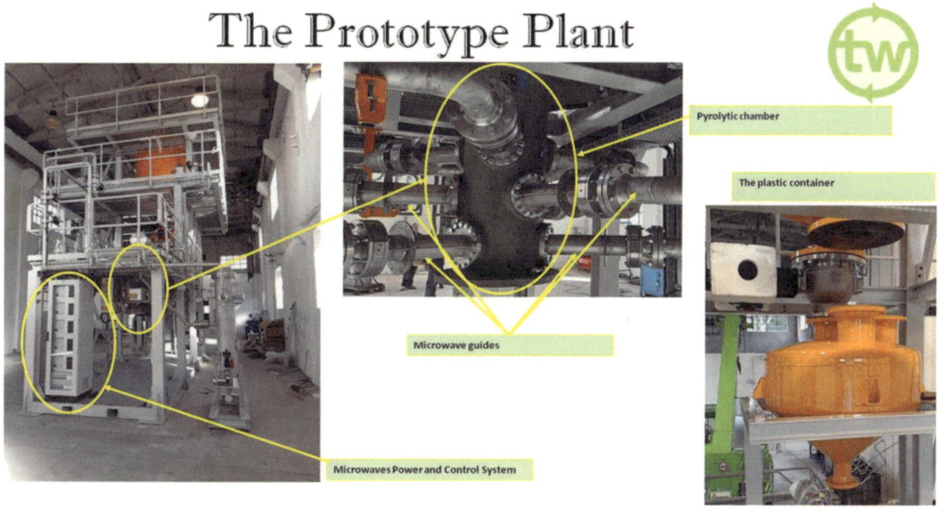


**Figure 1.**
*Some pictures of the prototype plant of Techwave s.r.l. (by curtesy of Techwave s.r.l.).*

The total emission was 25.6 MW, and this oven was able to pyrolyze 5 full ELT (almost 50 Kg) every cycle (30 min) (**Figure 1**). Other plants were reported by the Scandinavian Biofuel Company (SBC) [110], EWI, Molectra, Techwave [111], Pyrowave, and very recently by Resynergi [112]. Finally, there were several proposed industrial plants in the course of their realization, among which those proposed by Tyrebirth, Techwave, EWI, Pyrowave, or Resynergi.

a. Environmental Waste International - Canada (EWI): Located in Whitby, Ontario (Canada), since 1992, supplied proprietary systems for eco-friendly waste reduction, using patented reverse polymerization processes. These plants have been able to recycle scrap ELT since 1994, demonstrating their capabilities. Following their results, EWI realized a continuous feed system, then EWI amplified their field of expertise with plants able to dispose of contaminated products.

b. Molectra technologies - Au: Realized MAP of ELT. Waste management review reports a speech on molectra technologies' John Dobozy about "his unique approach to tire recycling and the development of several niche products for wide-ranging customers and markets" [111]

c. Scandinavian biofuels company (SBC) reports MAP, as well suited process to recycle a variety of waste fractions such as tires, sewage sludge, agricultural waste, waste wood, electronic scrap, cables, plastic waste, etc.

d. Tyrebirth s.r.l. - Italy: Tyrebirth has developed, following proprietary patents [113–115] a very highly efficient recycling process that allows the creation of high-quality products through the MAP of ELT (**Figure 2**). The Tyrebirth process was carried out at reduced pressure to avoid any gas leaking in the atmosphere, and it produces solid, liquid, and gaseous products: (a) Solid - carbon black (rCB) - (40% of tire treated) having high performances because it is dry,

**Figure 2.**
*A picture of the Tyrebirth plant (by curtesy of Tyrebirth s.r.l.).*

unpolluted and with quality characteristics 95% close to virgin carbon black. The rCB can be directly reused to produce new tyres, or employed as pigments, fillers for plastics, in textile printing, toner, and so on. (b) Solid - harmonic steel - (10% of ELT) easily separated and delivered to steelwork. (c) Liquid - SynOil with low sulfur content (< 1%) (35% of ELT). The SynOil is a low-sulfur diesel marine oil and may be employed in ships' diesel engines, in thermal plants for electricity production, and in other specific installations. (d) The gas (15% of ELT) contains liquid petroleum gas (LPG) and hydrogen. It may be used to produce more than 70% of the plant's electric energy required through cogeneration. The capacity of the plant is 540 Kg/h.

e. Techwave s.r.l. - Italy: This Italian start-up, founded in 2018, following an innovative technological solution and using proprietary patent [58, 116], has developed a MAP process to realize an industrial plant to recover materials and energy from waste plastic (**Figures 1** and **3**). The process is innovative, energetically sustainable, and has a very low environmental impact. The innovation of the process is the use of a microwave to supply the energy for plastic pyrolysis using an MW absorber, obtaining uniform heating of the material and avoiding the problems of classical thermal heating. Furthermore, the process is carried out at a reduced pressure, avoiding any dispersion of the gas in the atmosphere. Finally, another characteristic of the plant is its compactness because it may be installed in two standard containers.

f. Pyrowave - Canada: This company claims a MAP process [117], to regenerate post-consumer plastics by breaking into products useful to produce new plastics, restoring their full value. Pyrowave paves the way to a truly circular economy of plastics. The regeneration of plastics allows for infinite recycling and avoids the Green greenhouse gases associated with the extraction of virgin material.

**Figure 3.**
*Part of Techwave for gas purification, if required. (By courtesy of Techwave).*

Recently, they extended their activity and proposed the recovery of carbon fiber present in WTB or aircraft fuselages.

g. Resynergi - USA: This group initially patented a "Distillation system and method using MAP" [118] to obtain hydrocarbons from MAP of soapstock, followed by another patent on "Microwave system for converting Hydrocarbon Base Waste Materials into oil and fuel" [119]. A proprietary Resynergi modules was offered (**Figure 4**), and these modules are supplied, serviced, and warrantied by innovation partner Lummus Technology. The Resynergi modules utilize the Continuous Microwave Assisted Pyrolysis (CMAP) Technology engineered by Lummus, offering a decentralized, cost-effective solution to plastic recycling. Even if in the presentation the MAP of polyethylene and polypropylene is only reported, in the Resinergy presentation, "a groundbreaking technology makes easy the transformation of hard-to-recycle plastic waste into high quality circular pyrolysis oil to accelerate the circular economy" is reported.

**Figure 4.**
*The Resynergi reactor and a picture where some modules are linked together (by courtesy of Resynergi).*

## 4. Conclusion

This chapter reports many MAP applications where hazardous/contaminated waste plastics were friendly and economically disposed of, reducing the carbon footprint and obtaining valuable products useful to produce new plastic or other very interesting chemical compounds, giving a strong contribution to the realization of a circular economy.

The proposal of MAP is a very interesting way because all products obtained (char, oil, and gas) may be reutilized, so no waste is obtained from chemical recycling.

The process is self-sustainable from an energetic point of view, rendering the process economically sustainable.

The interest of industrial companies in this field is improving day by day, as reported in the last part of the chapter, because this is an important step for the realization of a truly circular economy and the reduction of the greenhousegas in the atmosphere.

## Author details

Marco Frediani[1*], Piero Frediani[1], Irene Mellone[2], Gianni Innocenti[2], Gianpaolo Oteri[2] and Roberto Simoni[2]

1 Department of Chemistry, Sesto Fiorentino - Florence, Italy

2 Cognito Engineering SRL, Firenze, Italy

*Address all correspondence to: marco.frediani@unifi.it

## References


[1] Farag S, Sobhy A, Akyel C, Doucet J, Chaouki J. Temperature profile prediction within selected materials heated by microwaves at 2.45 GHz. Applied Thermal Engineering. 2012;**36**:360-369

[2] Menéndez JA, Juàrez-Pérez EJ, Ruisànchez E, Bermùdez JM, Arenillas A. Ball lightning plasma and plasma arc formation during the microwave heating of carbons. Carbon. 2011;**49**:339012 36:360

[3] Undri A, Meini S, Rosi L, Frediani M, Frediani P. Microwave pyrolysis of polymeric materials: Waste tires treatment and characterization of the value added products. Journal of Analytical and Applied Pyrolysis. 2013;**103**:149

[4] Bykov YV, Rybakov KI, Semenov VE. High-temperature microwave processing of materials. Journal of Physics D: Applied Physics. 2001;**34**:R55

[5] Russell AD, Antreou EI, Lam SS, Ludlow-Palafox C, Chase HA. Microwave assisted pyrolysis of HDPE using an activated carbon bed. RSC Advances. 2012;**2**:6756

[6] Mohajerani A, Burnett L, Smith JV, Markovski S, Rodwell G, Rahman MT, et al. Recycling waste rubber tyres in construction materials and associated environmental considerations: A review. Resources, Conservation and Recycling. 2020;**155**:104679. DOI: 10.1016/j.resconrec.2020.104679

[7] Zafar S. Production and Applications of Crumb Rubber [Internet]. EcoMENA; 2018. Available from: https://www.ecomena.org/crumb-rubber-production-and-uses/

[8] End of Life Tyres: A Valuable Resource with Growing Potential. European Tyre and Rubber Manufacturers' Association; 2011. Archived from the Original on 23 November 2015 [Accessed: November 11, 2016]

[9] Sienkiewicz M, Kucinska-Lipka J, Janik H, Balas A. Progress in used tyres management in the European Union: A review. Waste Management. 2012;**32**:1742-1751

[10] "Stockpile Cleanup". Rubber Manufacturers Association. 15/03/2016. Archived from the Original on 17/11/2016. Retrieved 17 November 2016 – via Google Scholar. Available from: https://en.wikipedia.org/wiki/Tire_recycling [Accessed: July 03, 2023]

[11] Parkash S. Petroleum Fuels Manufacturing Handbook. New York: The McGraw-Hill Companies, Inc.; 2010

[12] Myhre M, Saiwari S, Dierkes W, Noordermeer J. Rubber recycling: Chemistry, processing, and applications. Rubber Chemistry and Technology. 2012;**85**:408-449

[13] Rodriguez IM, Laresgoiti MF, Cabrero MA, Torres A, Chomòn MJ, Caballero B. Pyrolysis of scrap tyres. Fuel Processing Technology. 2001;**2**:9

[14] Menéndez JA, Domínguez A, Inguanzo M, Pis JJ. Microwave-induced drying, pyrolysis and gasification (MWDPG) of sewage sludge: Vitrification of the solid residue. Journal of Analytical and Applied Pyrolysis. 2005;**74**:406

[15] Predel H. Ullmann's Encyclopedia of Industrial Chemistry. Karlsruhe, Germany: Wiley-VCH Verlag GmbH & Co. KGaA; 2000

[16] Murillo R, Aylón E, Navarro MV, Callén MS, Aranda A, Mastral AM. The application of thermal processes to valorise waste Tyre. Fuel Processing Technology. 2006;**87**:143

[17] Juma M, Korenová Z, Markos J, Annus J, Jelemensky L. Pyrolysis and combustion of scrap tire. Petroleum and Coal. 2006;**48**:15

[18] Crelling JC, Hagemann HW, Sauter DH, Ramani RV, Vogt W, Leininger D, et al. Coal. In: Ullmann's Encyclopedia of Industrial Chemistry. Weinheim, Germany: Wiley-VCH Verlag GmbH & Co. KGaA; 2000

[19] Chan OS, Cheung WH, McKay G. Preparation and characterisation of demineralised Tyre derived activated carbon. Carbon. 2011;**49**:4674

[20] Aylòn E, Fernàndez-Colino A, Navarro MV, Murillo R, Garcìa T, Mastral AM. Waste tire pyrolysis: Comparison between fixed bed reactor and moving bed reactor. Industrial and Engineering Chemistry Research. 2008;**47**:4029

[21] López FA, Centeno TA, Alguacil FJ, Lobato B, López-Delgado A, Fermoso J. Gasification of the char derived from distillation of granulated scrap tyres. Waste Management. 2012;**32**:743

[22] Ucar S, Karagoz S, Ozkan AR, Yanik Y. Evaluation of two different scrap tires as hydrocarbon source by pyrolysis. Fuel. 2005;**84**:1884

[23] Napoli A, Soudais Y, Lecomte D, Castillo S. Scap Tyre pyrolysis: Are the effluents valuable products? Journal of Analytical and Applied Pyrolysis. 1997;**40/41**:373

[24] Sahouli B, Blacher S, Brouers F, Sobry R, Van den Bossche G, Diez D, et al. Surface morphology of commercial carbon blacks and carbon blacks from pyrolysis of used tyres by small-angle X-ray scattering. Carbon. 1996;**34**:633

[25] Undri A, Sacchi B, Cantisani E, Toccafondi N, Rosi L, Frediani M, et al. Carbon from microwave assisted pyrolysis of waste tires. Journal of Analytical and Applied Pyrolysis. 2013;**104**:396-404. DOI: 10.1016/j.jaap.2013.06.006

[26] Darmstadt D, Roy C, Kaliaguine S. Characterization of pyrolytic carbon blacks from commercial tire pyrolysis plants. Carbon. 1995;**33**:1449

[27] Lam SS, Chase HA. A review on waste to energy processes using microwave pyrolysis. Energies. 2012;**5**:4209

[28] Horikoshi S, Matsubara A, Takayama S, Sato M, Sakai F, Kajitani M, et al. Characterization of microwave effects on metal-oxide materials:Zinc oxide and titanium dioxide. Applied Catalysis. B, Environmental. 2009;**91**:362

[29] Menéndez JA, Arenillas A, Fidalgo B, Fernández Y, Zubizarreta L, Calvo EG, et al. Microwave heating processes involving carbon materials. Fuel Processing Technology. 2010;**91**:1

[30] Piskorz J, Majerski P, Radlein D, Wik T, Scott DS. Recovery of carbon black from scrap rubber. Energy & Fuels. 1999;**13**:544

[31] Zainal NFA, Halim SIA. A review on demineralization, activation, and potential applications of the solid fraction obtained from the Pyrolyzed waste tires. Cleaner Waste Systems. 2025:100285. DOI: 10.1016/j.clwas.2025.100285

[32] Carbon Black Tire Grades: Sustainable Opportunities for the Manufacturing Industry. Available from: https://Contec.tech/carbon-black-grades-sustainable-alternatives/ [Accessed: July 18, 2023]

[33] United Nations. Sustainable Development Goals [Internet]. United Nations; 2025. Available from: https://sdgs.un.org/

[34] Scandinavian Enviro Systems [Internet]. Enviro; 2016. Available from: https://news.cision.com/scandinavian-enviro-systems/r/eco-friendly-rubber-with-recycled-carbon-black-now-in-volvo-production,c9906408

[35] Hull R. Volvo's Concept Recharge is Its Vision for Super-Green EVs of the Future [Internet]. This is Money; 2021. Available from: https://www.thisismoney.co.uk/money/cars/article-10212487/Volvos-Concept-Recharge-vision-super-green-EVs-future.html [Accessed: August 1, 2025]

[36] Michelin and Bridgestone See Potential for Demand for up to 1 Million Tons of Recovered Carbon Black in 2030 [Internet]. Envirosystems; 2022. Available from: https://envirosystems.se/mfn_news/michelin-and-bridgestone-see-potential-for-demand-for-up-to-1-million-tons-of-recovered-carbon-black-in-2030/ [Accessed: August 1, 2025]

[37] Pirelli Launches The P Zero E: The Sporting Tire that's A|Pirelli [Internet]. Pirelli.com; 2023. Available from: https://www.pirelli.com/tires/en-us/car/press-releases/pzero-e-230713

[38] Shao J, Tian X, Fan T, Zhang J, Wang S, Cheng W, et al. Study on waste tires pyrolysis characteristics and high-value oil: Effects of pyrolysis parameters and catalyst. Journal of the Energy Institute. 2025;**118**:101934. DOI: 10.1016/j.joei.2024.101934

[39] CORDIS cordis.europa.eu. Taking SynOil to the Market: Synthetic Oil (SynOil) from Plastic Waste with the Molten Zinc Reactor [Internet]. CORDIS|European Commission; 2022. Available from: https://cordis.europa.eu/project/id/884176/reporting [Accessed: August 1, 2025]

[40] BASF Signs Long-Term Agreement with Braven Environmental for Supply of Advanced Recycled Feedstock [Internet]. Basf.com; 2025. Available from: https://www.basf.com/us/en/media/news-releases/2025/03/P-US-25-06 [Retrived: April 20, 2025]

[41] Polimerica.it. Poliolefine da riciclo chimico [Internet]. Polimerica.it; 2021. Available from: https://www.polimerica.it/articolo.asp?id=25831 [Accessed: August 1, 2025]

[42] Polimerica.it. Poliolefine da riciclo chimico [Internet]. Polimerica.it; 2021. Available from: https://www.polimerica.it/articolo_pdf.asp?id=25403 [Accessed: August 1, 2025]

[43] Polimerica.it. Fondi UE per l'olio di pirolisi di OMV [Internet]. Polimerica.it; 2025. Available from: https://www.polimerica.it/articolo_pdf.asp?id=33504 [Accessed: August 1, 2025]

[44] Polimerica.it. Dow investe nel riciclo chimico in Europa [Internet]. Polimerica.it; 2025. Available from: https://www.polimerica.it/articolo.asp?id=33483 [Accessed: August 1, 2025]

[45] Polimerica.it. Prime consegne di PyOil a Shell [Internet]. Polimerica.it; 2025. Available from: https://www.polimerica.it/articolo.asp?id=33761 [Accessed: August 1, 2025]

[46] WasteExpo [Internet]. Wasteexpo.com; 2025. Available from: https://www.wasteexpo.com/en/home.html

[47] Polimerica.it. Pneumatici "green" al 70% [Internet]. Polimerica.it; 2025. Available from: https://www.polimerica.it/articolo.asp?id=33797 [Accessed: August 1, 2025]

[48] Wang S, Cheng M, Xie M, Yang Y, Liu T, Zhou T, et al. From waste to energy: Comprehensive understanding of the thermal-chemical utilization techniques for waste tire recycling. Renewable and Sustainable Energy Reviews. 2025;**211**:115354. DOI: 10.1016/j.rser.2025.115354

[49] Plastics Europe. Plastics – The Fast Facts 2024 • Plastics Europe [Internet]. Plastics Europe; 2024. Available from: https://plasticseurope.org/knowledge-hub/plastics-the-fast-facts-2024/

[50] The Plastic Waste Management Framework|AEPW [Internet]. Endplasticwaste.org; 2023. Available from: https://www.endplasticwaste.org/insights/reports/plastic-waste-management-framework

[51] Undri A, Rosi L, Frediani M, Frediani P. Microwave pyrolysis of polymeric materials. In: Chandra U, editor. Microwave Heating. London: IntechOpen; 2011. pp. 207-226

[52] Motasemi F, Afzal MT. A review on the microwave-assisted pyrolysis technique. Renewable and Sustainable Energy Reviews. 2013;**28**:317-330. DOI: 10.1016/j.rser.2013.08.008

[53] Kim J-K, Park K-B, Kim DK, Song S-H, Kim J-S. Direct production of olefins from waste plastic using a pyrolysis and fluid catalytic cracking integrated process: Part 1. Study on the production and analysis of waxy oil obtained using a new type of a fluidized bed reactor. Energy. 2024;**306**:132420. DOI: 10.1016/j.energy.2024.132420

[54] Undri A, Rosi L, Frediani M, Frediani P. Efficient disposal of waste Polyolefins through microwave assisted pyrolysis. Fuel. 2014;**116**:662-671. DOI: 10.1016/j.fuel.2013.08.037

[55] Bartoli M, Rosi L, Frediani M, Undri A, Frediani P. Depolymerization of polystyrene at reduced pressure through a microwave assisted pyrolysis. Journal of Analytical and Applied Pyrolysis. 2015;**113**:281-287. DOI: 10.1016/j.jaap.2015.01.026

[56] Undri A, Frediani M, Rosi L, Frediani P. Reverse polymerization of waste polystyrene through microwave assisted pyrolysis. Journal of Analytical and Applied Pyrolysis. 2014;**105**:35-42. DOI: 10.1016/j.jaap.2013.10.001

[57] Undri A, Rosi L, Frediani M, Frediani P. Conversion of poly(lactic acid) to lactide via microwave assisted pyrolysis. Journal of Analytical and Applied Pyrolysis. 2014;**110**:55-65. DOI: 10.1016/j.jaap.2014.08.003

[58] Frediani P, Rosi L, Frediani M, Undri A, Occhialini S. Production of Hydrocarbons from Copyrolysis of Plastic and Tire Material with Microwave Heating, WO2012110990, 2012, PCT/IB2012/050747

[59] Available from: https://www.eea.europa.eu/help/glossary/eea-glossary/european-waste-catalogue-1 [Accessed: April 17, 2025]

[60] Available from: https://en.wikipedia.org/wiki/Municipal_solid_waste#cite_note-3 [Accessed: May 17, 2025]

[61] Available from: https://www.eesi.org/files/eesi_msw_issuebrief_072109.pdf [Accessed: April 20, 2025]

[62] Agaton C, Boongaling G, Charmaine Samala V, Resy Ordona V, Riza O. Economic analysis of waste-to-energy investment in the Philippines: A real options approach. Applied Energy. 2020;**275**:115265. DOI: 10.1016/j.apenergy.2020.115265

[63] Rosi L, Frediani M, Undri A, Bartoli M, Frediani P. In: Tzortzakis N, editor. A Friendly Management of Waste/Contaminated Polymeric Materials from Differentiated Waste Collection through Microwave Pyrolysis, in Municipal Solid Waste. Novapublisher; 2017

[64] Ma M-Y, Xu D-H, Guo Y, Wang S-Z, He Y-L. Impacts of microplastics in municipal sludge pyrolysis-gasification: Melt-induced microaggregation and increased deposition risk. Renewable and Sustainable Energy Reviews. 2025;**214**:115542. DOI: 10.1016/j.rser.2025115542

[65] Wei S, Zhou T, Yang D, Yang Q, Luo S, Feng Y. Microwave-induced and catalytic pyrolysis of municipal domestic waste under multiple synergistic effects. Journal of Renewable and Sustainable Energy. 2025;**17**:025034. DOI: 10.1063/5.0250134

[66] Du H, Yuan D, Li W, Wang L, Li Y, Che L, et al. Efficient removal of toxic organics and reduction of Cr(VI) to Cr(III) from tannery sludge: A comparative study of microwave pyrolysis and conventional pyrolysis. Separation and Purification Technology. 2025;**354**:128736. DOI: 10.1016/j.seppur.2024.128736

[67] Anastas P, Warner J. 12 Principles of Green Chemistry [Internet]. American Chemical Society; 2024. Available from: https://www.acs.org/green-chemistry-sustainability/principles/12-principles-of-green-chemistry.html

[68] Techwave Private Communication. 2020

[69] Available from: https://www.fraunhofer.de/en/press/research-news/2021/june-2021/closed-loop-recycling-pilot-project-for-single-use-face-masks.html [Accessed: May 17, 2025]

[70] Available from: https://www.ewi.ca/ [Accessed: April 20, 2025]

[71] Per- and Polyfluoroalkyl Substances (PFAS). Washington, D.C.: U.S. Environmental Protection Agency (EPA); 2025

[72] Reconciling terminology of the universe of per- and Polyfluoroalkyl substances: Recommendations and practical guidance. In: OECD Series on Risk Management of Chemicals. OECD; 2021. p. 23. DOI: 10.1787/e458e796-en

[73] Geddes L. What are PFAS? Everything You Need to Know About the 'Forever Chemicals' Surrounding Us Every Day. London: The Guardian; 2024. Available from: https://www.theguardian.com/environment/article/2024/may/25/what-are-pfas-everything-you-need-to-know-about-the-forever-chemicals-surrounding-us-every-day [Accessed: July 31, 2025]

[74] Can We Take the 'Forever' Out of Forever Chemicals? www.bbc.com; 2023. Available from https://www.bbc.com/future/article/20231016-cleaning-up-pfas-forever-chemicals [Accessed: July 31, 2025]

[75] Schymanski EL, Zhang J, Thiessen PA, Chirsir P, Kondic T, Bolton EE. Per- and polyfluoroalkyl substances (PFAS) in PubChem: 7 million and growing. Environmental Science Technology. 2023;**57**:16918-16928. DOI: 10.1021/acs.est.3c04855

[76] PFAS Explained. EPA; 2024. Available from: https://www.epa.gov/

pfas/pfas-explained [Accessed: July 31, 2025]

[77] Kluger J. All the stuff in your home that might contain PFAS 'forever chemicals'. Time Magazine. 2023

[78] "PFAS and Your Health". Centers for Disease Control and Prevention. [Accessed: December 12, 2025]

[79] Perkins T. Toxic 'Forever Chemicals' Widespread in Top Makeup Brands, Study Finds, The Guardian, 15/06/2021. Archived from the Original on 7 July 2021

[80] Whitehead HD, Venier M, Wu Y, Eastman E, Urbanik S, Diamond ML, et al. Fluorinated compounds in north American cosmetics. Environmental Science & Technology Letters. 2021;**2021**(8):538-544. DOI: 10.1021/acs.estlett.1c00240

[81] US EPA. Per- and Polyfluoroalkyl Substances (PFAS) [Internet]. www.epa.gov; 2024. Available from: https://www.epa.gov/sdwa/and-polyfluoroalkyl-substances-pfas

[82] European Chemicals Agency. Perfluoroalkyl Chemicals (PFAS) - ECHA [Internet]. echa.europa.eu; 2024. Available from: https://echa.europa.eu/hot-topics/perfluoroalkyl-chemicals-pfas

[83] Bamdad H, Papari S, Moreside E, Berruti F. High-temperature pyrolysis for elimination of per- and Polyfluoroalkyl substances (PFAS) from biosolids. PRO. 2022;**10**:2187

[84] Hušek M, Semerád J, Skoblia S, et al. Removal of per- and polyfluoroalkyl substances and organic fluorine from sewage sludge and sea sand by pyrolysis. Biochar. 2024;**6**:31

[85] Zhang D, Song Q, Hou B, Zhang M, Teng D, Zhang Y, et al. Experimental study on microwave pyrolysis of decommissioned wind turbine blades based on silicon carbide absorbents. PRO. 2024;**12**:1065. DOI: 10.3390/pr12061065

[86] Åkesson D, Foltynowicz Z, Christéen J, Skrifvars M. Microwave pyrolysis as a method of recycling glass fibre from used blades of wind turbines. Journal of Reinforced Plastics and Composites. 2012;**31**:1136-1142. DOI: 10.1177/0731684412453512

[87] Åkesson D, Krishnamoorthi R, Foltynowicz Z, Christéen J, Kalantar A, Skrifvars M. Glass fibres recovered by microwave pyrolysis as a reinforcement for polypropylene. Polymers and Polymer Composites. 2013;**21**:333-340. DOI: 10.1177/096739111302100601

[88] Xu M, Ji H, Wu Y, Meng X, Di J, Yang J, et al. Recovering glass fibers from waste wind turbine blades: Recycling methods, fiber properties, and potential utilization. Renewable and Sustainable Energy Reviews. 2024;**202**:114690. DOI: 10.1016/j.rser.2024.114690

[89] Jasinska D, Dutkiewicz M. Waste Management of Wind Turbine Blades-a Review of recycling methods and applications in cementitious composites. Sustainability. 2025;**17**:805. DOI: 10.3390/su17030805

[90] Emmerich R, Kuppinger J. Recovering carbon Fibers. Kunststoffe International. 2014. DOI: 10.24406/publica-fhg-236236

[91] Available from: https://www.meccanicanews.com/2025/04/30 [Accessed: May 8, 2025]

[92] Yousef S, Eimontas J, Stasiulaitiene I, Zakarauskas K, Striūgas N. Recovery of energy and carbon fibre from wind turbine blades waste (carbon fibre/unsaturated polyester resin) using

pyrolysis process and its life-cycle assessment. Environmental Research. 2024;**245**:118016. DOI: 10.1016/j.envres.2023.118016

[93] Yao Y, Cao Y, Rao M, Yang J, Zhang Y, Gu J, et al. Production of green phenol by microwave-assisted catalytic pyrolysis of epoxy-based carbon fiber reinforced plastic from wind turbine blades. Journal of the Energy Institute. 2025;**119**:102005. DOI: 10.1016/j.joei.2025.102005

[94] Chin K-Y, Shiue A, You J-L, Wu Y-J, Cheng K-Y, Chang S-M, et al. Carbon Fiber recycling from waste CFRPs via microwave pyrolysis: Gas emissions monitoring and mechanical properties of recovered carbon Fiber. Fibers. 2024;**12**:106. DOI: 10.3390/fib12120106

[95] Dega C, Boukhili R, Esmaeili B, Laviolette J-P, Doucet J, Decaens J. Microwave-assisted pyrolysis of carbon Fiber-reinforced polymers and optimization using the box–Behnken response surface methodology tool. Materials. 2024;**17**:3256. DOI: 10.3390/ma17133256

[96] Hao S, He L, Liu J, Liu Y, Rudd C, Liu X. Recovery of carbon fibre from waste Prepreg via microwave pyrolysis. Polymers. 2021;**13**:1231. DOI: 10.3390/polym13081231

[97] Li Y-F, Li J-Y, Syu J-Y, Yang T-H, Chang S-M, Shen M-Y. Mechanical Behaviors of microwave-assisted pyrolysis recycled carbon Fiber-reinforced concrete with early-strength cement. Materials. 2023;**16**:1507. DOI: 10.3390/ma16041507

[98] Andersson M, Knutson Wedel M, Forsgren C, Christeen J. Microwave assisted pyrolysis of residual fractions of waste electrical and electronics equipment. Mining and Engineering. 2012;**29**:105-111. DOI: 10.1016/j.mineng.2011.09.005

[99] Rosi L, Bartoli M, Frediani M. Microwave assisted pyrolysis of halogenated plastics recovered from waste computers. Waste Management. 2018;**73**:511-522. DOI: 10.1016/j.wasman.2017.04.037

[100] Suriapparao DV, Batchu SP, Jayasurya S, Vinu R. Selective production of phenolics from waste printed circuit boards via microwave assisted pyrolysis. Journal of Cleaner Production. 2018;**197**:525-533. DOI: 10.1016/j.jclepro.2018.06.203

[101] Liu C, Li C, Shu B, XiaH LD. Preliminary strategy and product analysis of microwave pyrolysis of waste printed circuit board. Canadian Journal of Chemical Engineering. 2025;**103**:543-551. DOI: 10.1002/cjce.25387

[102] Zhang T, Mao X, Siyal Asif A, Fu J, Jiang Z, Deng Z, et al. Microwave-assisted catalytic pyrolysis of waste printed circuit boards, and migration and distribution of bromine. Journal of Hazardous Materials. 2021:402123749

[103] Andooz A, Eqbalpour M, Kowsari E, Ramakrisma S, Cheshmeh ZA. A comprehensive review on pyrolysis from the circular economy point of view and its environmental and social effects. Journal of Clener Productions. 2023;**388**:136021. DOI: 10.1016/j.clepro.2023.136021

[104] Huang Y-F, Cheng T-T, Chiueh P-T, Lo S-L. Resource recovery from spent crystalline-silicon solar modules by using microwave pyrolysis, acid leaching and chemical etching. Process Safety and Environmental Protection. 2025;**194**:47-57. DOI: 10.1016/j.psep.2024.12.017

[105] Huang Y-F, Chen Y, Chiueh P-T, Lo S-L. Metal recovery from copper indium gallium selenide solar cells by using microwave pyrolysis, thermal oxidation and thermal chlorination. Process Safety and Environmental Protection. 2024;**190 A**:226-232. DOI: 10.1016/j.psep.2024.07.042

[106] Frediani P, Frediani M. Private Communications

[107] Pyrowave. Pyrowave: A New Generation of Plastics [Internet]. Pyrowave. Available from: https://www.pyrowave.com/

[108] Available from: https://wastemanagementreview.com.au/a-rubberised-archetype-molectra-technologies/ [Accessed: April 20, 2025]

[109] Available from: https://www.tyrebirth.com/en/ [Accessed: April 20, 2025]

[110] SBC. Available from: https://www.sbiofuel.com/downloads.html [Accessed: April 20, 2025]

[111] Available from: https://www.techwave.it/ [Accessed: April 20, 2025]

[112] Resynergi - News [Internet]. Resynergi.com; 2025. Available from: https://www.resynergi.com/news [Accessed: August 1, 2025]

[113] Frediani P, Rosi L, Frediani M, Meini S, Undri A, et al. Produzione di idrocarburi da Pirolisi di gomme, CAF scarl, Italian Patent 1404132B1 (2011); Pat request N. FI2011A000030 - rif. N&G n. 10445PTIT (2011)

[114] Frediani P, Rosi L, Frediani M, Undri A, Occhialini S, Meini S. Production of Hydrocarbons from Pyrolysis of Tyres. WO2012110991 2012, PCT/IB2012/050748

[115] Innocenti G, Simoni R, D'Alessandro G, Andreini V, Frediani P, Sonego A, et al. Dispositivo di pirolisi, It Patent n. 102018000003163 (01/03/2018), ref. P020275IT-01/sf

[116] Frediani P, Rosi L, Frediani M, Undri A, Occhialini S. Produzione di idrocarburi da Co-Pirolisi di Materiali Plastici e Pneumatici con riscaldamento a Microonde, CAF scarl, Italian Pat 1404131B1 (2011); Patent Request FI2011A000029 - rif. N&G n. 10772PTIT (2011)

[117] Douchet J, Chaouki J, Sobhy A. Catalyst for Distributed Batch Microwave Pyrolysis, System and Process Thereof, WO2012/097448 A1 and Reference Therein Cited

[118] Rongsheng R, Anderson E. Distillation System and Method Using Microwave-Assisted Pyrolysis, US Pat 12,006,472 B2, Jun 11, 2024

[119] Bauer B, Tanne J, Spott T, Mayhew F, Mcinnis AJ, Leis M, et al. Microwave System for Converting Hydrocarbon Base Waste Materials into Oil and Fuel, EP 3 813 996 B1, 28.09.2022